STRIKE WARFARE
in the 21st Century

Titles in the Series

THE U.S. NAVAL INSTITUTE
BLUE & GOLD PROFESSIONAL LIBRARY

For more than 100 years, U.S. Navy professionals have counted on specialized books published by the Naval Institute Press to prepare them for their responsibilities as they advance in their careers and to serve as ready references and refreshers when needed. From the days of coal-fired battleships to the era of unmanned aerial vehicles and laser weaponry, such perennials as *The Bluejacket's Manual* and the *Watch Officer's Guide* have guided generations of Sailors through the complex challenges of naval service. As these books are updated and new ones are added to the list, they will carry the distinctive mark of the Blue and Gold Professional Library series to remind and reassure their users that they have been prepared by naval professionals and meet the exacting standards that Sailors have long expected from the U.S. Naval Institute.

STRIKE WARFARE
in the 21st Century

AN INTRODUCTION TO NON-NUCLEAR ATTACK BY AIR AND SEA

DALE E. KNUTSEN

NAVAL INSTITUTE PRESS
Annapolis, Maryland

Naval Institute Press
291 Wood Road
Annapolis, MD 21402

Library of Congress Cataloging-in-Publication Data
Knutsen, Dale E.
  Strike warfare in the 21st century / Dale E. Knutsen.
    p. cm.
  Includes index.
  ISBN 978-1-61251-083-5 (hbk. : alk. paper) — ISBN 978-1-61251-096-5 (ebook)
1. Air warfare. 2. United States. Navy—Weapons systems. 3. Air weapons. I. Title.
  UG633K58 2012
  359.40973—dc23

                                                        2011050977

19 18 17 16 15 14 13 12   9 8 7 6 5 4 3 2 1
First printing

Contents

Preface

Military operations are often the subject of adventure novels and documentary works, and there are numerous publications on the subject written for military or technical professionals. The works of fiction and the historical recounting of combat operations can make for interesting and exciting reading. The more technical documents are a great source of detail for the factors that influence the conduct and outcome of combat operations. But everyday citizens who would simply like to know more about the basics are often left in an awkward information void, caught between historic or fictional stories on the one hand and detailed discussions full of technical terms, jargon, abbreviations, and acronyms on the other hand.

This book attempts to bridge that chasm by dealing with complex topics in an organized fashion using everyday language. It is intended to eliminate much of the mystery that surrounds reports of our modern military operations. To reach this goal, this book begins with basics explained with simplified language and illustrations. Terminology, details, and necessary complications are gradually added, but always with a focus on fundamentals.

As an introduction to the subject, the material provides an overview of the many aspects of strike warfare and strike weapons. Readers who are familiar with the subject will probably find much of the material to be greatly simplified and lacking in depth. However, the book is intended for newcomers to the field, members of the news media, and the interested general public. Information is intentionally presented in small pieces to avoid overloading or intimidating the average citizen. A college degree or a military background is not required to understand the material presented here; readers only need to have curiosity about the subject. The material is presented in two parts. First is an overview of strike warfare operations, dealing with the background and activities associated with combat actions. This part relates most closely to what is presented in the news media. The second part of the book deals with the development and acquisition of strike weapons, a subject that is less newsworthy but has a large influence on how strike warfare is ultimately carried out.

Several appendixes are found at the end of the book. These contain summarized or additional information, and may satisfy those seeking somewhat greater technical depth. As with any collection of data, the information reflects what was known at the time it was published. In the case of strike warfare and strike weapons, that matter faces the added complication of security constraints; for obvious reasons, only unclassified data will be found in this book.

This project could not have been completed without the patience and support of my life partner and wife of fifty years, Georgia. Not only did she provide moral support, she also served as the first reviewer of the text, and tactfully offered suggestions in many areas that were initially too technical, confusing, or downright boring. I also need to express sincere thanks for review inputs provided by a select group of colleagues: Gerald O. Miller, Sterling Haaland, James Knepshield, Monte Frisbee, and D. Keith Crummer. Their insights kept me from drifting too far off course in several areas.

My goal was to help Americans better understand some of the tools and factors that influence military operations during times of tension or conflict. If readers gain a better appreciation of those factors through this book, that goal will have been achieved.

PART 1
Strike Warfare Operations

ONE

Introduction to Part 1

Disagreements between groups of human beings probably began not long after groups of human beings first appeared on this planet. Sometimes those disagreements escalated to the point of violence. When the violence persisted beyond a brief exchange, the period of hostilities became known as "war."

The history of mankind has been punctuated with frequent periods of warfare. Armed clashes and major campaigns are memorialized in records that began in the dim past. While little may have been recorded about everyday life in ancient times, significant efforts were made to record battles and victories in carvings, monuments, and texts. It is as if the ancients were measuring their worldly importance in terms of their success on the battlefield. Or perhaps it was an early form of intimidation aimed at neighbors: "Don't mess with us—we're tough."

In more recent times, warfare has expanded in terms of numbers of combatants, geographic scale, and methods of combat. This was most evident during the twentieth century when larger and larger numbers of warriors from multiple nations fought each other with increasingly deadly weapons. The twentieth century also saw the advent of an entirely new branch of combat through the introduction of the airplane and later the guided missile.

When we reduce conflict to its most basic level, war has been and remains a cruel, bloody, and costly endeavor. Wars are fought over the disagreements of the time and with the weapons of the time, and continue until one of the antagonists defeats the other or until one or both of them determine that it is too painful or costly to continue. To those who die in battle, it matters little whether the mortal wound is from a club, a sword, a bullet, or an explosion; the fatal outcome is the same.

Wars occur when diplomacy fails. History is full of examples of many different kinds of diplomatic failures that led to armed conflict, sometimes in spite of sincere efforts of one side to reach a peaceful accord. The reasons for conflict are generally described in economic, religious, ethnic, or personal-liberty terms, but the underlying roots often seem to be material greed and an obsession with power. Since greed and a desire for power appear to be unfortunately common human traits, this bodes poorly for any real change in how future groups of humans treat each other.

In spite of this rather bleak overview of the role of warfare in human society, the latter half of the twentieth century did see some significant advances being made in reducing unwanted destruction and casualties in the war zone, an issue commonly called collateral damage. Much of the reduction in collateral damage can be attributed to technological improvements that led to greatly increased accuracy of firepower. Simply put, it became easier to put destructive force on a specific target with a limited number of shots, rather than having to employ a barrage of shots to level an entire area. Much of this advancement is found in the evolution of strike warfare.

Strike Warfare Defined

In the broadest definition, "strike warfare" is the use of aircraft, ships, or submarines to employ nonnuclear weapons against targets on the land or sea surface. Some military organizations prefer the phrase "tactical strike warfare" to emphasize the nonnuclear, or "conventional," role. Others restrict "strike warfare" to targets on land, using the term "antisurface warfare" to apply to ship targets. However, for the purposes of this book, "strike warfare" will address both land and sea targets.

There are a few peculiarities buried within this definition. In common practice, strike warfare focuses on various kinds of weapons that are fired, released, or launched only from aircraft, ships, or submarines. It thus excludes weapons that might be employed by infantry or other ground units. It also excludes torpedoes fired by submarines or ships, and does not include sea mines that are intended to create a barrier against ships or submarines.

The very wide array of strike warfare targets will be explored in the next chapter, but suffice to say that they can be just about anything, fixed or moving, that is found on land or the surface of the sea. Another branch of conflict, known as "antisubmarine warfare," or ASW in military jargon, deals with submerged submarine targets, while yet another, called "antiair warfare," or AAW, focuses on airborne targets such as aircraft and cruise missiles.

Within these broad, target-oriented definitions the military adds several additional subgroupings that tend to focus on particular kinds of operations or circumstances. Some of those will be discussed in later chapters.

A Brief History

It could be argued that the earliest form of strike warfare was the use of cannon fire from sailing ships to bombard other ships or targets near the shoreline. What began with muzzle-loading guns that were aimed visually gradually transitioned over several centuries to today's stabilized, rapid-fire cannon that make use of sophisticated aiming and fire control systems. Naval gunfire remains an important element in the mix of shipboard weapons, but it is not the primary focus of this book for the simple reason that it has become more of a defensive tool to help protect the ships from attack. At the present time, circumstances seldom favor naval gunfire in a primary offensive role against other combat vessels or against land targets farther inland than the shoreline area. Instead, guided missiles and weapons launched from aircraft have become the dominant tools of strike warfare. But it took a substantial evolution of hardware and tactics to bring us to the present situation.

World War I saw the introduction of the aircraft as an offensive tool of war. The earliest bombers were few in number and carried a very limited load of small bombs. While the military impact of these bombers was relatively small, their psychological and strategic impacts were large. Areas behind the lines could no longer be considered completely safe. Even cities far from the front could be attacked with bombs dropped from zeppelins.

Damage from World War I bombers was relatively light, mostly due to poor delivery accuracy and the small bombs employed. However, something fundamental had changed in the scope of warfare, and the flying machine had made a deadly impression on the world.

By the time hostilities began in World War II, aerial bombing had evolved into a major factor in warfare. Bombers were carrying heavier loads and optical bombsights were being used to improve delivery accuracy. Nevertheless, it was still common practice to drop multiple bombs against even a small target in hopes that one of them might land close enough to destroy the object. "Carpet bombing" was widely employed against industrial areas or other large targets, and "strings" or "sticks" of bombs were released to increase the probability of success. When bombing precision was essential, dive-bombers or very close, low-altitude releases were generally required; in both of these situations, the aircraft delivering the bombs became quite vulnerable to antiaircraft fire.

Germany introduced unmanned guided weapons during the latter part of World War II. The V-1 "buzz bomb" is considered to be the first operational cruise missile. The V-1 carried a large explosive charge in a conventional-looking, twenty-six-foot-long airframe that was powered by a simple pulse jet engine. It was rail launched from a fixed-ground installation and then flew a compass heading to a specified distance, up to about 140 miles, where it then dived onto the target area and detonated. V-1s flew at fairly low altitudes (two thousand to three thousand feet) at speeds that were comparable to some of the swifter fighter aircraft, making them vulnerable to antiaircraft defenses and even fighter attack. The very noisy pulse jet engine could be heard at long distances and helped alert air defenses to inbound missiles. Because of its limited accuracy, the V-1 was used primarily against large industrial complexes or cities.

More dramatic was the German A-4, or V-2, ballistic missile. This was a sleek, forty-five-foot-tall rocket that was launched vertically from a simple mobile pad. The warhead of the V-2 was only slightly heavier than that of the V-1, and its range was only about 30 percent greater, but the V-2 most certainly had a greater psychological impact, arriving without warning over the target at supersonic speeds. The rocket engine in the V-2 burned kerosene and liquid oxygen, propelling the missile high into the atmosphere before it plunged down steeply on the target area. A very rudimentary form of inertial guidance was used, which again resulted in relatively poor accuracy by today's standards.

Other forms of guided weapons appeared during the latter stages of World War II, exhibiting varying degrees of success. Some were attempts to remotely control explosive-laden aircraft, while others were experiments in purpose-built guided missiles. However, none but the German V-1 and V-2 were employed in substantial numbers.

Guided weapons captured the imagination of military strategists on both sides of the Iron Curtain during the Cold War. Much of the attention in the United States and the USSR was focused on ways of delivering nuclear warheads over long distances, to supplement their large forces of strategic bombers carrying nuclear bombs. As ballistic missile technology matured, there first appeared intermediate range ballistic missiles (IRBMs), then intercontinental ballistic missiles (ICBMs), and finally the smaller, submarine-launched fleet ballistic missiles (FBMs).

Long-range strategic cruise missiles were also explored and produced. Early experiments with intercontinental-range cruise missiles were largely pushed aside by the success of the ballistic missile. But later programs such as the American Tomahawk and the air-launched cruise missile (ALCM) fielded large numbers of strategic weapons.

On the nonnuclear, or "tactical," side, hard experience during the Korean conflict led to increased interest in improving what were known as "conventional" weapons. A variety of issues had emerged by the mid-1950s, including matters involving weapon compatibility with new generations of aircraft, concerns about aircraft survivability as enemy air defenses improved, and a need for greater overall delivery accuracy. These three fundamental issues, expressed in various ways, drove the evolution of strike weapons for the next half century.

The Emergence of Standoff, Precision Strike

Over the course of history, there has been a recurring pattern of weapons development. When one nation develops a new weapon, opposing nations generally attempt to do two things in response: (1) they attempt to develop their own version of the new weapon, and (2) they try to develop a means of defeating their opponent's new capability. The natural response to the appearance of warplanes in the early twentieth century was (1) the introduction of opposing warplanes and (2) the introduction of antiaircraft guns.

With time, defenders became more proficient with their steadily improving antiaircraft weapons. Attacking airplanes became increasingly vulnerable if they flew very close to a defended target to drop their bombs. As a result, attackers attempted to release their bombs farther away from the target to increase their own chances of survival. However, the crude bomb-aiming techniques initially in use resulted in very poor bomb accuracy from these greater distances. This then led to more work developing improvements to bomb aiming from higher altitudes. And so on.

When antiaircraft defenses are minimal, attacking aircraft can make a very close approach to the target before releasing their weapons and not place themselves at high risk. But if there are significant antiaircraft defenses in place, such a close approach would likely result in the aircraft being destroyed. Under those more dangerous circumstances, the attacker would do well to employ a longer-range, or "standoff," weapon that would allow weapon release from the aircraft at a relatively safe location.

The concept of standoff is not new. In the earliest of human conflicts, the individual who carried a rock in his hand had an advantage over a bare-fisted opponent, because the one who threw a rock could do so from a safer position. Likewise, when flint blades were introduced, the one who fastened a stone point to a stick to fashion a spear had an advantage over the one who only had a knife. Standoff allowed the individual to fight from a safer position. And so it is today.

The problem with standoff is that it becomes more difficult to be accurate as you move away from the target. If you can jab your opponent with a spear from just over arm's length away, you can probably be fairly accurate. But if you must throw the spear at the target from some distance away, accuracy suffers.

Attaining both standoff and accuracy has been a continuing challenge for decades. High-altitude bombing in World War II relied on high-quality optical bombsights to reduce errors, but bombs still had to be dropped in large numbers to ensure that at least some of them arrived at the desired aimpoint. Atmospheric uncertainties and winds, minor perturbations and errors during release, and small variations in the ballistics of the bombs themselves all contributed to a substantial amount of dispersion by the time the bombs reached the surface.

Adding some kind of guidance to a bomb seemed like a way of reducing errors, thereby requiring fewer bombs (and bombers) to destroy a surface target. That notion proved to be correct, but it took several design iterations, starting with what are referred to as "dumb bombs," before arriving at today's array of guided weapons, often called "smart weapons." Achieving this goal has been made easier over time as guidance components have become more precise, less bulky, and lighter.

As will be described in later chapters, a standoff precision strike weapon is far more complicated and expensive than an unguided bomb. It will also become clear that there are important applications for unguided bombs, long-range precision strike weapons, and the items in between. The challenge is in arriving at an appropriate mix of weapons.

Unmanned Operations

The most recent major change in strike warfare has been the use of unmanned, or uninhabited, vehicles in both surveillance and attack roles. The philosophical roots of this change can be traced to the tendency in Western society to place a high value on the lives of its warriors, which then leads to significant measures to protect those warriors from harm or death. The use of body armor, defensive support tactics, and standoff weapons are all examples of measures intended to reduce combat risk.

When it became both technologically feasible and operationally practical to remove human beings from some types of combat vehicles, the military began to exploit that capability in high-risk or long-duration missions. The initial and still most widespread application is associated with airborne surveillance, whereby unmanned air vehicles (UAVs) carry various kinds of sensors aloft to observe objects and activities in the combat zone. Armed UAVs then followed, which allowed direct attack of surface targets without exposing a human pilot to air defenses. Further discussion of these developments will be found later in the book, but it is clear that a significant change has taken place in the basic conduct of strike warfare.

TWO

Targets

In the broadest sense, anything that is found on the surface of the land or the sea could be considered a potential strike warfare target. Structures, facilities, roads, rail lines, bridges, stored supplies, ships, vehicles, troops—all are possible targets during time of conflict. Many of these items are considered to be "civilian" in nature, but that matters little during wartime, when the only real issue is whether destruction of the item will impede the enemy's ability to continue the conflict.

Fundamental Characteristics

Strike warfare targets exhibit an exceptionally wide range of characteristics. They can be as large as a major industrial factory or as small as a squad of troops. They can be as "hard" as a buried bunker or as "soft" as an exposed aircraft. And they can be fixed and motionless, such as a building; movable but motionless, such as a mobile missile launcher at rest; or in motion, such as an armored vehicle on land or a ship at sea.

If we look first at size, we can find targets varying from a few square feet to tens of thousands of square feet in area. All of the objects are three-dimensional, but they vary significantly in configuration: some are relatively flat while others are relatively tall, with the majority falling somewhere in between. These variations may seem mundane to most people, but to weapons designers and military operators who must be able to neutralize such objects, the variations are quite significant.

The relative "hardness" of a target is another serious consideration. Objects that are relatively "soft" can be easily damaged or destroyed by the effects of an explosive blast or high-velocity fragments, even if the strike weapon does not directly hit the object. "Hard" targets, on the other hand, may require both a direct impact and a specially designed warhead to inflict major damage.

Fixed targets, particularly those that are sizable, are easier to locate and attack than objects that can be moved or are in motion. An airfield on land, for example, is easier to find than an aircraft carrier at sea. Similarly, a military command center in a building is easier to attack than a mobile command trailer that is moved from day to day to different areas of the battlefield. It is also obvious that it is simpler to hit a stationary building than a moving vehicle.

As will be seen in a later discussion, these three fundamental characteristics— size, hardness, and motion—are critical considerations when the time comes to attack a particular surface target.

Target Settings

Another important consideration in strike operations is the setting and background of the target. This factor has to do with the relative ease or difficulty of detecting and identifying a target, and then directing a weapon to it. An isolated building on an open plain poses an easier problem than a building in mountainous, forested terrain, or that same building among many other buildings in an urban setting. The current weather and lighting at the target also play a role. Finding the target during the day in clear weather is typically easier than finding it on a stormy night.

There usually are limits on weapon and operational capabilities based on these kinds of factors. The basic setting may create difficulties due to terrain, elevation, foliage, adjacent objects (often called "clutter"), and general background. The local environment at the time of the attack will also influence the operation, based on lighting (day, twilight, or night), general climate (prevailing humidity, seasonal conditions) and current weather (fog, clouds, precipitation, and wind). Ideally we would prefer to have weapon systems available that are usable anywhere in the world, during any season and weather, and at any time. For a variety of reasons, mostly related to affordability and practical operational issues, we often fall short of that goal.

Target Complications

A third major factor that must be considered has to do with coincidental or intentional target complications. These may arise when the target is intermingled with or close to civilian (noncombatant) objects, or when the enemy has intentionally used camouflage, concealment, or deception measures to "hide" the target. The appearance of the target can be manipulated through a variety of techniques, including measures to reduce its observable "signature," such that it becomes difficult to detect or identify as the intended object of the attack. More will be said about such complications in a later chapter dealing with the surveillance portion of strike operations.

Generic Surface Targets List

Even if we simplify the strike warfare target question by ignoring the issues of target settings and complications, a very complex list of target types still emerges. Target lists are popular with the people in the Pentagon who have the responsibility of defining weapon system requirements. A target list allows the requirement to be focused on specific concerns and to frame the evaluation procedure that will be used during weapon development. If a weapon is needed to destroy enemy radars, for example, the requirement statement that is issued by the Pentagon will identify the types of radars to be addressed, and the development program will include evaluations of the new weapon against those kinds of radars.

At a very basic level, surface targets can be separated into either land or sea categories, and then further separated into broad subsets of objects. The lists that follow use an arbitrary set of categories to organize a number of generic (nonspecific) surface target types, expressed in common military terms. In actual practice target lists tend to become classified when they identify specific targets, such as particular types of armored vehicles, radars, ships, and so on.

Land Targets
Major Offensive Sites
- Airfield elements: aircraft on the ground, runways, hangars, aircraft shelters/bunkers, fuel and munitions storage, support buildings
- Surface-to-surface missile (SSM) sites: fixed or relocatable launchers, missiles, support vehicles and buildings

MAJOR DEFENSIVE SITES

- Surface-to-air missile (SAM) sites: launchers, missiles, launch control facilities
- Electronic sites: radars, electronic receiver facilities, transmitters and repeaters
- Command, control, and communications (C3) sites: associated buildings, bunkers, and mobile units; related links and information processing/distribution sites

BATTLEFIELD AND FORWARD AREA SITES

- Tactical defenses: mobile point defense and close-in SAMs, antiaircraft artillery (AAA)
- Vehicles: heavy tanks, armored fighting vehicles (AFVs), armored personnel carriers (APCs), self-propelled and towed artillery, trucks, smaller vehicles (e.g., the ubiquitous Toyota pickups driven by insurgents)
- Field fortifications: improvised or permanent emplacements, bunkers
- Troops: in foxholes or trenches, in the open

LOGISTICS, SUPPORT, AND OTHER TARGETS

- Lines of communication/commerce (LOC): bridges, roads, railroads, rail yards, rolling stock, trucks
- Port facilities: piers, docks, wharfs, dry docks, shipyards, warehouses, unloading cranes
- Buildings: military and industrial structures of all types
- Power generation facilities: power plants, penstocks, transformer yards, transmission lines, dams
- Storage facilities: petroleum tank farms, ammunition dumps, warehouses

SEA TARGETS

WARSHIPS

- Combatants of all types: aircraft carriers, cruisers, destroyers, frigates, patrol craft

SUPPORT AND MERCHANT SHIPS

- All types: tankers, freighters, underway replenishment vessels, landing craft, utility craft

OTHER

- Unconventional vessels: small, high-speed smugglers' craft, semisubmersible craft

As noted earlier, submerged submarines are targets for another branch of warfare logically titled antisubmarine warfare (ASW). Since modern submarines are seldom found on the surface of the sea, it would be unlikely to find them on a strike weapon target list. However, surfaced submarines in port or in dry dock would be valid strike targets even though they are rarely listed.

As you can now appreciate, a major challenge in strike warfare is arriving at a mixed inventory of strike weapons that can cope with this very broad array of target characteristics without becoming overly cumbersome or unaffordable.

THREE

Defenses

No worthwhile target goes undefended in time of conflict. The nature and extent of such defenses can vary widely, but defenses of some kind can be anticipated for every significant object on a target list. The challenge for attackers is to first determine what kind of defense exists and then attempt to counter it.

Passive Defenses

There are some situations in which the very location or setting of a target acts as a defense against attack. For example, distance alone is sometimes considered to be an adequate defense if a target is thought to be located beyond the practical range of attacking forces. Modern long-range strike weapons have tended to reduce the reliance on the distance-only defense, but its application continues to influence defense strategy. This is especially true for the United States, which has for decades used the presence of friendly neighbors to our north and south and of the Pacific and Atlantic Oceans to our west and east as a passive defense against easy attack. What American policy documents refer to as a "forward strategy" simply means that it has been and remains U.S. policy to engage the enemy as far away from American soil as possible. As a result, we no longer find many surface-to-air missile (SAM) sites or interceptor aircraft bases within the continental United States even though they were prevalent in the 1950s.

Camouflage has been used since early times to make it difficult to find targets. Initially the emphasis was on visual camouflage, because historically most attacks were based on visually observing the object. Irregular paint schemes that mimic the background, netting, and attachment of foliage were all used to cause the object to blend into the background. As other types of sensors began to be used to locate targets, it became necessary to try to expand camouflage techniques to include ways of obscuring, altering, or masking the target from infrared sensors and radars as well.

Decoys may be used in conjunction with camouflage measures to deceive the attacker. In this instance the actual target is camouflaged or sufficiently altered in appearance that it no longer matches the object that the attacker is searching for, but a short distance away the defender creates a false target that does match the sought-after characteristics. In the heat of battle, the attacker will "find" the decoy, and even though it isn't exactly where the target was thought to be located, it is "close enough" to the anticipated location and will therefore cause the real target to be missed.

Decoys may also be used independently to alter the attacker's perception of the type, location, or extent of the defender's forces. Phony facilities, vehicles, missile sites, and the like have all been used with varying degrees of success in causing the attacker to divert resources or alter battle plans.

Another form of a passive defense is found in some target settings. Attacking a railroad bridge target in open terrain is less of a challenge than attacking a similar bridge in a twisting and deep canyon. The attacker's task is also made difficult if the target is protected by robust structures, such as heavy concrete, earthen embankments, or similar materials. Aircraft shelters made of reinforced concrete with rock and earth added on top are a prime example of what can be considered a "hardened" target.

Perhaps the most extreme example of a passive defense is found in the target that is completely and deeply underground. Because of the effort and expense involved, this kind of defense is generally reserved for only the most valuable or critical facilities. In relatively open country, this usually involves considerable initial earthmoving to uncover a site for the facility. This is followed by sturdy construction and then reburial of the facility, perhaps with rock rubble added to the dirt above the structure. Alternatively it may involve tunneling into suitable cliffs, hills, or mountains to create a labyrinth of passages and interior spaces. Deeply buried facilities are generally considered "very hard" targets.

Active, Nonlethal Defenses

Some forms of defense involve actions taken during an attack that are expected to improve the survivability of the target while not intending to shoot down attacking aircraft or missiles. Perhaps the earliest such technique was the use of smoke to obscure the attacker's view of the target area. Smoke, however, is subject to the whims of the atmosphere; if the wind is blowing the smoke in the wrong direction, there is little the defender can do. Smoke also has little effect on attacking systems that use radar or other nonvisual types of sensors.

Some strike systems make use of infrared (IR) sensors on the launch aircraft, the strike weapon, or both. A clever defender may therefore attempt to mask or alter the target's infrared appearance through the use of techniques to selectively cool parts of the target, create artificial hot spots, or saturate the IR sensor with a large heat source. Alternatively a particularly valuable target might be protected with a laser device that is intended to "blind" an incoming IR-guided weapon and thus cause it to miss the target.

As more and more strike weapon systems make use of GPS (Global Positioning System) guidance techniques, defenders are hard at work developing ways of selectively degrading or negating GPS performance. GPS-jamming devices are intended to block out, corrupt, or spoof GPS signals in an area around the jammer, causing strike weapon accuracy to suffer. However, GPS jammers, or any kind of electronic jammer for that matter, must be used with care and discipline to avoid causing problems with equipment on both sides of the conflict.

Similarly, electronic devices are sometimes used to jam attackers' radars, cause errors in or inject false target returns into the attacker's radar, and so on. Use of such electronic countermeasures does bring with it the possibility that the attacker might instead return fire with a weapon that homes in on the jamming transmitter signal.

It should have become obvious by this point that anytime an overt action is taken to defend a target, it may draw attention to itself and may therefore become a "target of opportunity." The defender must exercise care when using such techniques, which work best only when the intention of the attacking force is clearly known. Turning on a jammer or a similar device at the wrong time could actually tip off the presence of a target and provide the attacker important tactical information that the defender would prefer to keep hidden.

Lethal Defenses

The final category of defenses includes those that are most familiar to people: guns, missiles, and airplanes. These are all designed to destroy the attacker's strike weapon systems, generally in a spectacular manner.

Effective employment of lethal defenses, and indeed any active defense measure, depends in large part on the ability of the defender to observe the presence and movement of the attacking forces. During the earliest airborne attacks, the defender relied almost exclusively on visual detection of inbound aircraft to sound the alarm. World War II saw the introduction of radar in some areas to provide a longer-range detection capability, but other methods continued to be used as well, including visual, sound, and radio transmission detection.

In more recent eras radar has dominated the surveillance and early warning field, but radar is definitely not the only method employed. Modern optical sensors search the sky for telltale infrared energy emitted by flying objects, and electronic sensors listen for transmissions from the attackers' radars, radios, electronic altimeters, and so on.

The individual bits of surveillance information obtained from these various sensors may not provide a complete picture of the attacker's activity. However, when all of the information is gathered together and displayed in a combined fashion, a much more comprehensive view is obtained. This is generally accomplished at a facility that then controls the response of the defenses in a coordinated manner. Such sites, generally called command-and-control centers, are very important to an effective air defense and are themselves high-priority targets from the attacker's point of view.

The defender's goal is to deploy a combined set of resources that includes the surveillance, early warning, and tracking sensors with various forms of lethal and nonlethal defenses that are employed against the attacking forces. The lethal defenses typically consist of:

- Fighters
- Surface-to-air missiles
- Antiaircraft artillery

Fighter aircraft equipped with air-to-air guided missiles and guns can fly out long distances to intercept attacking forces far from targets. The fighters would be directed (vectored) to the incoming attackers by the command-and-control center; historically this was called a ground-controlled intercept (GCI). As the fighters approach the intercept point, they use their own on-board sensors to detect the attackers and engage them with air-to-air missiles (AAMs). If the fight continues to the point that the aircraft are in very close proximity to each other, guns may eventually be employed.

Surface-to-air missiles (SAMs) are available in a wide array of capabilities. The longest-range SAMs are typically found with multiple items of associated equipment. The missiles themselves might be transported on a large vehicle designed to elevate them into a firing position. Nearby would be a local-surveillance radar and one or more fire control (FC) radars used to direct the missiles toward their assigned targets. Other vehicles would carry spare mis-siles and necessary support equipment.

Long-range SAMs often need capable onboard guidance sensors to pro-vide the necessary accuracy to hit an airborne target a long distance from the launch site. This level of sophistication adds complexity and cost to such weapons, making them the "silver bullets" of the defense world.

Medium- and short-range SAM systems use smaller missiles and tend to have fewer vehicles or associated pieces of equipment. Some incorporate all of the key functions into one tracked or wheeled vehicle and, except for the surveillance aspect, form a mobile, stand-alone battlefield air defense system. Both radar and optical fire control systems can be found in use. The very-shortest-range SAMs are the small, shoulder-fired missiles that are often called MANPADS (man-portable air defense system).

Antiaircraft artillery (AAA) comes in a variety of calibers and types, from the basic machine gun pointed into the sky, to the large-caliber gun firing a heavy shell equipped with a proximity or time fuze, to the multibarrel rapid-fire cannon directed by an associated radar. While some people might think that AAA has been replaced by modern guided missiles, it still fills an impor-tant low-altitude gap in the capabilities of other lethal defenses.

Lasers may ultimately be introduced on the battlefield in significant numbers to supplement and perhaps eventually replace AAA and short-range SAMs. Rather than relying on the impact of fragments, projectiles, or blast waves to inflict damage on the attacker, lasers use intense heat generated on a small area to destroy the structure or ignite the components of the inbound aircraft or missile.

Defense Envelope

The amount of sky protected by these various defenses is generally depicted in what is referred to as a "defense envelope." The most common form of envelope shows the altitude and range capability of the defense system (fig. 3.1). There are usually some assumptions or conditions that are printed on the diagram, explaining some important details that affect capabilities.

Interpretation of the envelope is fairly straightforward. The message is that the defense system is expected to be able to intercept and damage or destroy an inbound attacker anywhere within the area circumscribed by the envelope boundaries. The envelope defines the maximum-altitude capability, the maximum-range capability, the minimum-range capability, and so forth.

Each defense system has its own unique effectiveness envelope, and those envelopes do change when the underlying assumptions or conditions change. For example, basic SAM envelopes usually assume an inbound attacker that is easy to detect, is moving at a moderate speed, is moving directly toward the defense site, does not employ countermeasures, and does not maneuver (change direction or speed). Envelopes usually assume fair weather, no significant winds, and no terrain or man-made obstructions. All of that leads to

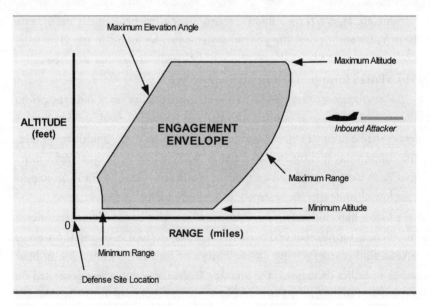

Figure 3.1 Generic Defense Envelope: Surface-to-Air Missile

performance estimates that tend to be optimistic in terms of what the defense system might be capable of under actual combat conditions. In short, they typically describe the best-case scenario for the defender.

Defenses are seldom used in isolation. In a given area of conflict, there may be zones of airspace protected by fighters, a very few long-range SAMs, a considerably larger number of medium- and short-range SAMs, and a very large number of AAA systems. The defense strategy will be to deploy those systems in a manner that results in overlapping fields of fire, avoiding holes in the defense, and providing some mutual protection. Surface targets that are especially valuable to the defender (logically called "high-value targets") will receive greater protection than objects of lesser concern.

The combined envelopes of these multiple defense systems can be illustrated as shown in figure 3.2. In this example we see a portion of the forward airspace allocated to fighter defense, often called a "fighter engagement zone," or FEZ. Behind that is airspace allocated to various surface-to-air missiles, also known as a "missile engagement zone," or MEZ.

The various systems in this example have been arranged to provide multiple layers of defense. Attackers that get by the fighters are then subject to

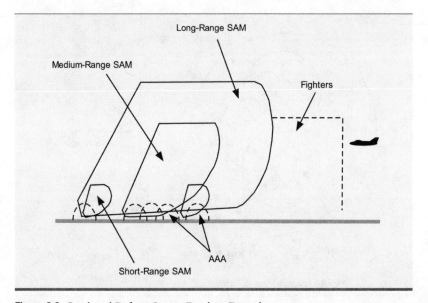

Figure 3.2 Combined Defense System Envelope Example

intercept first by long-range SAMs and then by medium-range SAMs. Low-flying attackers may additionally encounter short-range SAMs (including shoulder-fired missiles) and AAA. These latter systems are often intentionally arrayed to drive attackers up in altitude, where they are more vulnerable to medium- and long-range SAMs.

This kind of range-versus-altitude diagram is quite useful in understanding the extent of defenses in what is called the vertical plane, but it only provides a view along one particular axis or slice of the defended airspace. Another type of illustration places defense information in a "map view" of the area, indicating the geographic location of the various defense sites. An example of this kind of illustration is found in figure 3.3, in which the defended airspace is viewed from high above the region in what aviators often call a "God's-eye view." In this example the maximum engagement range of each of the systems is drawn as a circle around the site location.

It should be noted that both figures 3.2 and 3.3 are drawn in simplified fashion to avoid the complications that arise when the effects of terrain are added. These initial examples assume a completely smooth Earth with no significant terrain, obstructions, or battlefield complications.

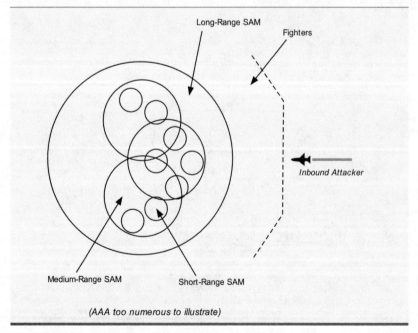

Figure 3.3 Defense Site Deployment: Map View Example

Effective employment of this varied assortment of defenses can only oc-cur when there is significant coordination between the many individual ele-ments. Such coordination usually relies on an array of radars and other detec-tion devices that are deployed in a manner intended to provide surveillance coverage of the region. The intent of the defender is to obtain both long-range and short-range coverage with as few gaps as practical. Information from the surveillance sites is routed to regional command-and-control facilities, where it is combined to form an overall picture of airborne events in the region. Decisions are then made and engagement orders are transmitted to the ap-propriate defensive sites.

On the land surface, long-range radars are often situated on elevated sites to minimize terrain obstructions. Nevertheless, rolling or mountainous ter-rain can still cause gaps in radar coverage, which leads to the practice of in-stalling shorter-range "gap-filler" radars in certain locations to fill those voids. Shorter-range radars are also found associated with mobile defense systems, and these may be used both for nearby surveillance purposes as well as for fire control purposes.

Airborne radars such as the American AWACS (airborne warning and control system) place the sensor high in the air, giving the radar an excellent opportunity to see inbound attackers at very long range. However, practical limits on size, weight, and power for radars carried in aircraft mean that air-borne radars usually cannot be as large as those found on the ground, which does place limits on the capabilities of airborne sensors. At the same time, the airborne platform allows the radar to "look down" into areas that would nor-mally be below the usable horizon for surface radars. It should also be noted that, due to their high cost, not all nations possess airborne radars.

Applying the same format as previously used for defense system enve-lopes, figures 3.4 and 3.5 illustrate the basic elevation and map envelopes for generic defensive radar systems. In the elevation example (fig. 3.4), the upper and lower extents of the detection envelope represent elevation limits while the outer extent represents the detection limit for the radar assuming standardized conditions. The elevation example does illustrate some funda-mental terrain issues, while the map example (fig. 3.5) again assumes no such obstructions.

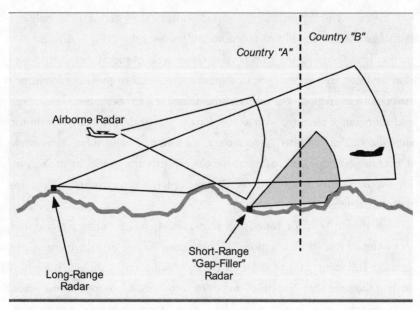

Figure 3.4 Radar Elevation Coverage Example

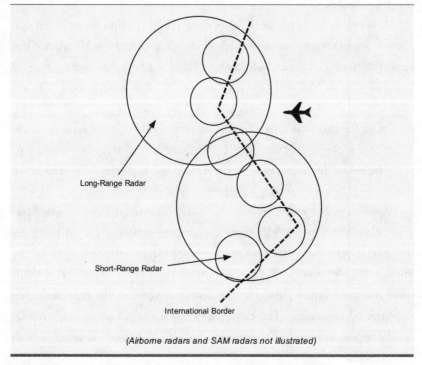

(Airborne radars and SAM radars not illustrated)

Figure 3.5 Radar Deployment Example

When all of these various surveillance, detection, and active defense elements are properly combined and coordinated, the defender is said to have an "integrated air defense system," or IADS, in military jargon. Accomplishing that combination and coordination usually requires a very large effort involving lines of communication (radio links, telephone lines, fiber-optic lines), computer systems and displays, appropriate facilities, and trained personnel. As might be expected, all of these elements, which are intended to defend against attack, then become targets of attack themselves. In fact, they are generally among the highest-priority targets during the initial phases of a conflict.

FOUR

Strike Resources

Having briefly explored potential enemy targets and target defenses, it is now appropriate to consider the tools available to the attacker. While the previous descriptions have been completely generic and not associated with any particular nation, the discussion of strike resources will follow the patterns developed by the United States. Similar capabilities also exist in other nations but may be identified using different terminology or organized in different arrangements.

Launch Platforms

In American military parlance, it is commonplace to use the term "launch platform" to identify the mobile object that fires, releases, or otherwise launches the actual strike weapon. The launch platform has historically been a manned vehicle of some sort, be it a ship, a submarine, or an aircraft. In more recent times, unmanned launch platforms have been introduced; the best-known examples of these have been battlefield drones, otherwise known as unmanned aerial vehicles (UAVs), which have been equipped with short-range air-to-surface weapons.

Launch platforms perform a variety of functions in the employment of strike weapons. In the simplest of cases, they may carry the strike weapon to a point in space and just release it. In more complex examples, the launch platform may use its sensors to locate the target, select a strike weapon for the mission, move the weapon onto an appropriate launch mechanism, check out the weapon and load mission data into it, aim the weapon toward the target, launch the weapon, and perhaps track or interact with the weapon on its inbound flight to the target.

SHIP AND SUBMARINE PLATFORMS

Historically, the earliest mobile launch platform for a strike weapon was a sailing ship equipped with muzzle-loading cannon. In the current era, ships continue to be important strike weapon launch platforms when they are equipped with surface-to-surface missiles (SSMs) in addition to their large-caliber guns. Modern combatants are typically outfitted with sophisticated sensors to locate targets in their vicinity, and are also able to make use of surveillance information received from other sources to attack targets at even longer range.

Small missile-equipped combatants such as patrol craft usually are armed with antiship missiles only, while larger vessels may carry a mix of ship attack and land attack strike weapons. In many cases the weapons are carried in sealed containers in an exterior area of the ship. Such containers protect the weapons from the elements, are mounted in an elevated firing position, and function as the launching mechanism. In other situations guided missiles are stowed belowdecks in magazines or in containers that also serve as launchers.

Submarines are also capable of launching strike weapons. While torpedoes remain a weapon of choice against surface ships at relatively close ranges in addition to their application against other submarines, strike weapons can extend their antiship capability out to longer ranges. Strike weapons also add a significant land attack role that historically was absent. Submarine-launched missiles may leave the underwater vessel from either horizontal torpedo tubes or vertical missile launch tubes, and must travel through the water as their first step on their way to the target.

The number of strike weapons carried by ships or submarines tends to be relatively limited due to space constraints on board the vessels. Once those missiles have been expended, the vessel must generally withdraw from the combat zone to replenish its strike weapon supply.

Ships can roam the seas with relative freedom, but they are obviously restricted to the water environment. Targets that are farther inland than the shipborne strike weapons can reach are thus safe from attack from these platforms.

AIRCRAFT PLATFORMS

Manned aircraft are a second category of strike weapon launch platforms. Most are fixed-wing aircraft, but there are some helicopters that carry short-range strike weapons. Aircraft may be based at land airfields and on aircraft carriers at sea, and range in size from large, long-range bombers to smaller, shorter-range fighter-bombers or attack aircraft.

Just as with ships, aircraft platforms are usually equipped with sensors that can be used to locate the target prior to weapon launch. Some aircraft are equipped with a laser that is used to shine infrared energy on a target and "designate" it for a laser-guided bomb or missile. Other aircraft-weapon combinations make use of an electronic link after launch to communicate between the weapon and the platform. Such an arrangement is called a "data link" (DL), and may involve the transmission of commands or revised target data to the weapon, transmission of in-flight information or images back from the weapon, or both.

Strike weapons may be carried externally on the aircraft or they may be carried in internal bomb bays. The latter approach is favored by large bombers and by radar-evading stealth aircraft. The strike weapon carriage capacity of an individual aircraft is quite limited, but aircraft do have the capability to quickly return to base and reload for a subsequent mission.

The unrefueled operating range or combat radius of strike aircraft imposes limits on their ability to reach distant targets. Such limits are relatively well known for fixed bases on land, but are a bit more flexible for those mobile bases afloat known as aircraft carriers. In both cases, however, in-flight refueling can greatly extend the reach of such aircraft.

UNMANNED PLATFORMS

The advent of unmanned aerial vehicles in the combat zone has altered some aspects of strike warfare. UAVs originally were used in surveillance and target location ("targeting") roles, but there has been a gradual increase in the use of armed UAVs to attack surface targets. It is important to note that while the UAV itself is not manned, current practice has a human controller monitoring the vehicle from a remote location and taking control of it when required. In particular both the decision and the command to launch a weapon from a UAV have thus far been under positive human control.

The initial application of armed UAVs has featured the use of small, light-weight weapons with relatively short range, carried under the wings of the air vehicle. The choice of weapons was partly due to the limited carriage capacity of early UAVs, and partly due to the need for positive control over their launch. However, the future application of unmanned platforms is not necessarily restricted to small weapons. Concepts in development for unmanned combat air system (UCAS)/unmanned combat air vehicle (UCAV) applications include carriage of weapons normally associated with larger manned aircraft. These latter applications typically feature internal carriage of the weapons.

OTHER PLATFORMS

It is fully recognized that there are a variety of surface-to-surface missiles launched from the land surface, but these systems are not covered in this particular book. Such systems include coastal defense missiles used against ships and tactical ballistic missiles such as the infamous Scud. Fixed installations are sometimes employed, but more often these are mobile weapons launched from towed trailers or from self-propelled transporter-erector-launchers (TELs).

Another conceptual platform that will not be addressed here is an unmanned armed vessel or submersible. There is no technological reason why a waterborne equivalent of a UAV could not be developed, but its place in military operational priorities remains uncertain.

SUMMARY

Launch platforms (fig. 4.1) come in a wide array of types, each with its own particular set of capabilities and limitations. As will be noted in part 2 of this book, there are specific platform compatibility factors that must be addressed in the design and use of weapons from each particular kind of platform. The common practice of developing weapons for a number of different platforms means that all of these combined compatibility requirements must be met by the weapon, often leading to design challenges and compromises.

In operational situations the availability of launch platforms plays a major role in combat planning. Their types, numbers, and locations all weigh heavily on how strikes will be conducted. Each launch platform is treated as a valuable asset, much more difficult to replace than the weapons it carries. Launch platform survivability is therefore a major consideration in strike planning.

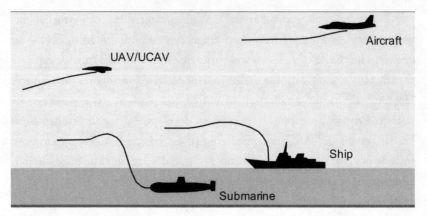

Figure 4.1 Generic Launch Platforms

Strike Weapons

The actual weapons used in strike operations are even more diverse than the launch platforms. At any given time the American strike weapons inventory will include items ranging in age from very recent to decades old, with a corresponding spread in technical sophistication. The combined inventory will include surface-to-surface missiles launched from the sea as well as air-to-surface missiles (ASMs) launched from aircraft. The weapons span the spectrum from very simple to exceedingly complex, vary in size from relatively small to quite large, in standoff from short to long range, and in accuracy from modest to precise. The basic reasons for some of these variations should become more apparent as you read on.

There are a number of terms used in the media to describe strike weapons: "bomb," "missile," "cruise missile," "laser-guided bomb" (LGB), "smart bomb," and so on. Sometimes the words are used accurately, correctly describing a particular form of weapon, while at other times the choice of wording is misleading. For purposes of this book, the generic term "weapon" is used to encompass the entire spectrum of destructive devices used in strike warfare.

At the most fundamental level, a bomb (fig. 4.2) is a relatively simple piece of ordnance made up of an explosive charge inside a metal shell, some kind of stabilizing device or fins at the rear end, mechanical attachments (bomb lugs) to allow it to be carried by an aircraft, a fuzing-and-safety device, and a small booster charge. The latter items are probably not familiar to most people and may need a little more explanation. The fuzing-and-safety device is intended to keep the bomb (or warhead) in a safe condition until it has been intentionally released or launched. After a suitable delay, it changes from a safe to an armed state and then awaits a firing signal from a target detector, which may simply sense impact. At that point a firing impulse is sent to the booster charge, which is used to set off the main explosive charge.

At the next level of sophistication, a guided bomb adds components needed to provide the necessary guidance signals, control fins to alter the flight path of the bomb, some kind of power supply, and perhaps larger lifting surfaces.

When substantial wings are added, you have a glide bomb or a glide weapon (fig. 4.3). This kind of weapon often confuses people who aren't quite sure whether to call it a guided missile. Add propulsion and the confusion goes away; it is definitely called a guided missile at this point.

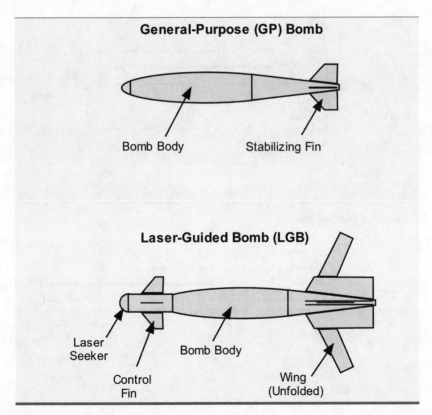

Figure 4.2 Notional Bomb and Laser-Guided Bomb

The semantics debate matters little in operational applications. What counts is the capability that the weapon brings to the fight, not what the item may be called.

When describing a strike weapon, it is common practice to first indicate its standoff category, then its launch platform type, and finally its target application, as, for example, "a short-range, air-launched, ship attack weapon." However, in operational planning, the more usual approach is to first identify the target to be attacked, then determine the needed standoff based on target defenses, and finally assess the availability of appropriate weapons from whatever launch platforms may be in the combat zone. Since this portion of the book is operationally oriented, the discussion of strike weapons will begin with target applications.

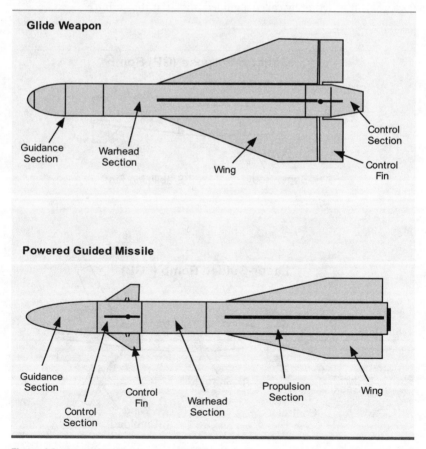

Figure 4.3 Notional Glide Weapon and Guided Missile

TARGET APPLICATIONS

The earlier discussion of target types should have made the point that their breadth of characteristics leads to a requirement for more than one type of attack weapon. Contrary to the hopes and desires of policy makers over a number a decades, there really is no single solution applicable to all of these target types.

When a target has been identified for a strike, one of the initial questions is "What will it take to inflict the required level of damage?" A logical first step in the process is therefore to consider the kinds of warheads or damage mechanisms that are available in the attacker's arsenal.

Warheads

The term "warhead" refers to the portion of the weapon whose primary function is to cause damage to the target. The most basic form of destructive device used against a surface target is a simple projectile that carries no explosive, as, for example, a solid cannonball or bullet. When such objects impact a structure, their kinetic energy causes portions of the structure to break, bend, or come apart. Kinetic energy projectiles are simple, inexpensive, and reliable. But their damage potential is limited by virtue of the requirement that they must physically impact some critical portion of the target, and their damage radius is very small. As a result simple projectiles have been almost universally replaced by various kinds of "warheads" that are intended to enlarge the damage radius and enhance their effects through the stored chemical energy of a high explosive. Kinetic energy is still used to help a warhead penetrate the target structure, but kinetic-energy damage effects are generally secondary to that of the primary explosive charge.

Historically, the term "warhead" has been applied to a destructive device that consists of an outer container, usually metal, filled with high explosive that is set off or detonated by some kind of fuzing mechanism, generally with a small boosting charge between the fuze and the explosive. The term "damage mechanism" refers to the various kinds of destructive forces produced by warheads. These include extreme pressures produced by a blast wave, high-velocity metal fragments from the outer case, and intense heat; these mechanisms are called blast, fragmentation, and incendiary effects.

The weapon solutions that have evolved with strike warfare display a history of attempts to be flexible in their application wherever feasible. Starting with simple bombs and large-caliber naval-gun projectiles, there are "general-purpose" designs that are intended for use against a broad array of objects, typically in the "soft" to "medium hardness" categories. Such targets are easily damaged by the blast and fragmentation effects of general-purpose (GP) warheads. The GP-warhead designs are relatively straightforward (fig. 4.4) and feature just enough metal casing to protect the explosive while penetrating light structures prior to detonation. GP warheads come in a range of sizes and weights, allowing strike planners to select the appropriate amount of explosive force for the particular target. For greatest destruction, the attacker generally attempts to place the warhead inside the target before detonation, thus confining the explosive effects to the most vulnerable target components. The destructive effects of GP warheads decline very rapidly if the warheads are detonated externally to or farther away from the target.

Sizable targets that are "hard" require a much more robust warhead. Such designs feature a thicker and heavier metal case, plus shock-resistant explosive and fuzing components, to allow the warhead to penetrate deep into earth or concrete before detonating. If the total warhead weight must be held to a fixed value, trade-offs must be made between explosive weight and case weight (and therefore target penetration capabilities). For "light" targets, explosive weight can be a large fraction of the total. But for "hard" targets, the necessary heavier case for penetration means that explosive weight is a smaller fraction of the total.

Some targets are particularly vulnerable to fire. Incendiary warheads or incendiary additives to GP warheads might then be selected for the mission. In most cases the incendiary materials are reactive solid materials that produce intense, concentrated heat for a sufficient period to ignite combustible portions of the target.

Other targets can be destroyed or compromised if a number of small projectiles impact them. This leads to fragmentation (frag) warheads, which are intended to send a spray of high-velocity fragments into such things as radar antennas, parked aircraft, electronic sites, and other relatively "soft" targets.

Armored vehicle targets such as tanks are something of a special case for airborne attack. They are relatively small, they may be moving, and their armor makes them virtually impervious to the effects of a near miss of even a very large GP warhead. However, they can be rendered inoperative by a small amount of explosive or fragmentation energy delivered inside their protective shell. This vulnerability led to the development of conical-shaped charge (CSC) warheads that are designed to concentrate a very large amount of energy on a small spot, which punches a hole in the armor plating and destroys the tank interior.

Another variation on focused energy is the linear-shaped charge (LSC) warhead. The lightweight metal shell of this kind of device resembles a shallow, multipoint star in cross section. Instead of directing the explosive energy forward on a single spot, the LSC warhead sends out multiple "lines" of energy along the axis of the warhead, slicing the structure of the target. The following blast wave then basically disassembles the weakened target.

Yet another explosive warhead type makes use of a blast-only or a blast-and-fire damage mechanism. The fuel-air explosive (FAE) design can produce exceptionally high peak pressures that can cause damage a substantial distance away from the detonation point in targets such as field fortifications, large buildings, or tunnels. The FAE warhead contains a combustible liquid or powder that must first be dispersed by an internal burster charge. After a very brief time interval to allow the material to mix with the surrounding air, the cloud is then detonated with a second charge, resulting in a violent explosion.

Some of the above damage mechanisms have been incorporated into miniature versions of the warheads weighing no more than a few pounds. These are called submunitions or bomblets, and are usually packaged into bomb-like containers ("dispensers") that are then dropped over the targets in the battlefield. At some point after being dropped from the aircraft, the dispenser opens, releasing the submunitions and allowing them to disperse and scatter, leading to a shotgun effect on the surface. This technique was originally developed as a way of using simple, unguided weapons to attack small, moving targets such as tanks or to distribute small warheads over a large target area.

The above-mentioned warheads are representative of what are considered "conventional" designs. In more recent times, some unconventional warheads or damage mechanisms have also been considered, tested, and developed for strike weapon usage. Very little public information has been officially released about such activities, but there have been sufficient brief commentaries in trade journals to provide some basics.

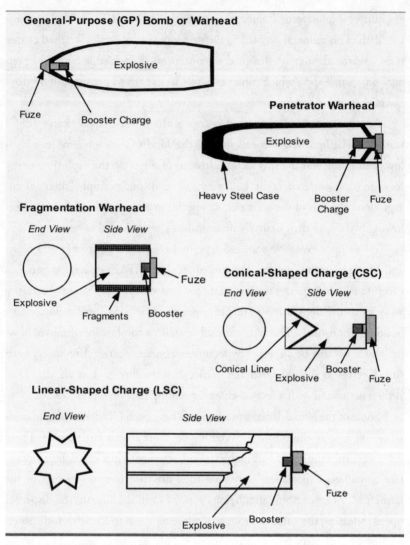

Figure 4.4 Notional Generic Warheads

One unconventional warhead that was allegedly first used against Iraqi electrical power distribution targets in 1991 was described as a container full of thousands of "conductive strings" carried by a cruise missile. When over the target, the container dispersed a cloud of these items, which floated down into transformer yards and related power lines, shorting out the electrical power grid.

More recent hints of warheads intended to disrupt or destroy electronic systems have focused on intense bursts of microwave energy near the targeted sites. The initial concept appears to involve either single-use or reusable micro-wave energy generators carried by guided weapons or UAVs to within close range of the designated site. At that point the high-power microwave (HPM) device would function, causing major damage to the targeted electronics. More recent concepts have suggested the use of microwave energy focused on the target site from a system on board a manned aircraft.

One way of summarizing the various forms and functions is found in figure 4.5, in which three different criteria are used to describe warhead types. The left column lists basic kinds of damage mechanisms, while the center column describes basic forms. The right column lists the more common descriptive terms used by the military services. Each column is relatively independent, meaning that there is no fixed correlation between terms. For example, there is a widely used cluster weapon that dispenses multipurpose submunitions that employ blast, fragmentation, incendiary, and conical-shaped charge damage mechanisms.

Finally, work apparently continues on the use of lasers from airborne plat-forms to disable or destroy objects on the surface. The most vulnerable targets would seem to be optical systems, but lightly constructed physical objects, such as exposed aircraft or missiles, could also be targets.

Guidance

Closely coupled with the issue of damage mechanism is the matter of accu-racy, which basically translates into weapon guidance or lack thereof. Early strike weapons, such as large-caliber gunfire and bombs, were completely un-guided and relied on whatever aiming or fire control systems were available to initially send them on their way toward the target. Once the projectile or weapon leaves the launch platform, there is no further intervention to correct its flight path. Weapons that are operated in this manner are said to be "ballistic." Many such weapons remain in use, largely because they are simple and inex-pensive. They do tend to produce considerable dispersion around the target (fig. 4.6) and usually require multiple shots or releases to achieve the desired level of target damage. Dispersion is a natural result of small errors in aiming, wind effects, slight differences in the bombs themselves, and a host of other "minor" factors.

Damage Mechanisms	Warhead Forms	Warhead Types
Kinetic Energy	Unitary Warhead[1]	Fuel-Air Explosive (FAE)
Blast	Submunitions[2]	High-Explosive (HE) Blast
Fragmentation	Unconventional: Multistage	Fragmentation (Frag)
Fire	Nonexplosive	General Purpose (GP)[3]
Electrical/Electronic Disruption		Incendiary
		Shaped Charge: Conical (CSC) Linear (LSC)
		Unconventional: Conductive Strings Energy Pulse

[1] That is, a single device.
[2] Also called cluster bombs or bomblets.
[3] These are generally blast-fragmentation devices.

Figure 4.5 Generic Warheads and Damage Mechanisms

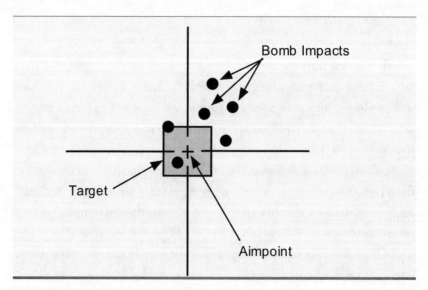

Figure 4.6 Example of Dispersion

Some of the earliest tactical guided weapons made use of "command guidance" to alter their flight path after launch. The pilot or other airborne operator observes the flight path of the inbound weapon and sends steering corrections to it by way of a radio, optical, wire, or fiber-optic data link. This type of manual guidance remains popular with short-range antitank weapons used on the battlefield but has generally fallen out of favor for strike weapons.

A more modern variation of manual guidance is found in what are generally called laser-guided bombs. Rather than tracking the inbound weapon, a human operator trains a spot of laser energy on the target and the weapon guides itself to that spot. An optical device called a laser designator is used to place the spot on the target and to keep it there until the weapon impacts. The laser designator can be on an aircraft, some other mobile platform, or even carried by a person on the ground. An optical receiver on the front of the weapon observes the reflected spot of laser energy and guides the weapon to the target by issuing steering commands to the weapon's control fins.

The guidance device generally found on the front of a guided weapon is called a "seeker" if it searches for and homes in on some specific kind of energy or target characteristic. If the weapon is launched without already having the seeker locked onto a specific target, it is called a "lock on after launch" (LOAL) system.

The laser seeker is only one of several kinds of devices used on strike weapons. Missiles that are intended to destroy radar sites use seekers that home in on transmissions of specific radar frequencies and are called antiradiation missiles, or ARMs. The ARM seeker is a "passive" receiver, meaning that it does not transmit. A more complicated radar guidance mode is found on antiship missiles equipped with "active" seekers that make use of small radar transmitters and receivers, allowing them to scan the ocean ahead of the missile, searching for the telltale reflection from a ship target.

Optical seekers other than laser devices are also in use. In their most basic form, they are aimed at and locked onto some portion of the target prior to launch; these are called "lock on before launch" (LOBL) systems. Once released from the launch platform, the seeker tracks that target feature and steers the weapon to it. The earliest strike weapon seekers in this category used black-and-white television (TV) tube technology. With time, imaging infrared (IIR) devices largely supplanted the television systems, adding a night capability that the TV devices could not provide. The IIR seekers are sensitive to thermal (heat) radiation differences and produce a picture similar to the forward-looking infrared (FLIR) systems that have become familiar to most citizens through news media coverage.

Seekers alone are generally not sufficient to provide complete guidance information to strike weapons, particularly those that must either fly at low altitudes or for a considerable distance. Some form of autopilot is usually incorporated into the weapon guidance system. This may be as simple as a means of keeping track of weapon orientation, so that steering commands can be properly transmitted to the control fins, or it may be as complex as a complete aircraft-type autopilot system. In the latter case this often includes a small inertial reference or inertial navigation system (INS) to keep track of weapon position in space, plus some form of barometric or radar altimeter.

With the advent of the Global Positioning System (GPS) satellites that provide worldwide location data, GPS receivers have become commonplace on a number of strike weapons. In most cases they are used to update and correct the weapon's inertial data, keeping the weapon on course over lengthy periods.

Combining a small inertial reference and a GPS receiver leads to a relatively simple guidance package that greatly improves the accuracy and operational flexibility of formerly unguided bombs. This guidance mode provides a predictable flight path to a set of target coordinates on the surface, regardless of small differences in ballistics or in-flight perturbations due to winds and weather.

Some strike weapons include a provision for postlaunch monitoring or intervention by human operators. This involves a communication link, generally called a data link, between the inbound weapon and either the launch platform or some other manned site, be it airborne or on the surface. In some examples the link only works in one direction, with the weapon transmitting a very brief data burst toward the end of the flight. That data burst might be as simple as a position and status message, or it might include an image of the target just before impact. Other data link examples provide communications in both directions, with information or images being sent by the weapon and steering updates being sent by the human operator. In this latter case it is possible to send a weapon on its way without having the seeker locked onto a target, and then to manually perform or confirm a target lock-on later in flight. This is sometimes called a "ready-shoot-aim" mode. Finally, a data link can also make it possible to revise the lock-on point as the weapon approaches the target, otherwise known as "sweetening the aimpoint."

Early tactical cruise missiles had no access to GPS technology and thus required some alternative form of navigation updates en route to the target. They used a terrain elevation correlation technique based on flying over pre-designated areas on the ground and comparing the sensed elevation profile (from a radar altimeter) with stored data. As they approached the target, a digital scene-matching correlation approach was used, comparing the observed optical image with a stored image. Both of these techniques required extraordinary prestrike surveillance effort and data manipulation to ensure reliable correlation. The great lengths that were required to obtain and process such data meant that this type of guidance was originally used in nuclear-armed strategic cruise missiles only.

Figure 4.7 summarizes the principal guidance modes that have been or are currently used in various strike weapons.

WEATHER ISSUES

These various guidance modes are not without their limitations. Each must be delivered to within an acceptable distance of the target in order to function properly. Many types of seekers require a clear line of sight to the target, meaning that there can't be obstructions such as terrain, foliage, or structures between the seeker and the target during the final approach.

Weather plays a very important role in the employment of

Guidance Modes
Unguided (Ballistic)
Command Guided:
Radio
Connected:
Wire
Fiber-Optic Cable
Preprogrammed:
Inertial (INS)
Inertial + GPS
Homing (Sensor/Seeker Equipped):
Laser Seeker (LGB)
TV Seeker
Infrared (IR) Seeker
Passive Radar Seeker (ARM)
Active Radar Seeker (Antiship)
Other:
Terrain Correlation (Radar Profile)
Scene Matching (Optical)
Two-Way (Radio) Data Link

Figure 4.7 Generic Guidance Modes

strike weapons. Clouds or fog in the target area can block the line of sight of optical modes such as television, imaging infrared, and laser seekers. Rain and snow can have a similar impact, and can reduce the effective range of radar seekers as well. High winds generally don't affect seekers but can perturb the strike weapon itself and can create dust clouds that decrease visibility for optical seekers. Similarly, dust from military operations or from previous strikes can pose problems for optical seekers.

Some weather effects are a bit more subtle. In tropical regions where warm temperatures and very high humidity are commonplace, the large amount of moisture in the air can reduce the effective range of imaging infrared devices. Atmospheric temperature and solar radiation can also influence the background setting of targets as viewed by imaging infrared seekers. At sea, the presence of high waves can reduce the effectiveness of radar seekers searching for ship targets.

All of these natural phenomena must be considered in the process of strike planning. Unsuitable en route or target area weather conditions are just cause for postponing a strike or for selection of an alternative attack strategy.

STANDOFF CATEGORIES

Some strike weapons must be launched very close to the target while others have the ability to fly very long distances. These differences in standoff capability are primarily used to help protect the launch platform from target defenses; short-range weapons can be used when defenses are light, while long-range weapons are usually needed when there are formidable defenses in depth. Longer standoff capabilities are also important to ship and submarine platforms that are inherently restricted to offshore launch positions.

As a matter of convenience, strike weapons are typically grouped by their basic standoff capabilities. While the terminology tends to change with time, one popular set of criteria arranges standoff groupings in terms of their ability to reach beyond certain classes of defenses, as follows:

- Standoff outside theater defense (SOTD): very-long-range capability, beyond the reach of the longest-range SAMs and generally outside the reach of enemy fighters
- Standoff outside area defense (SOAD): long-range capability, beyond the reach of intermediate-range "area" SAMs

- Standoff outside point defense (SOPD): medium-range capability, beyond the reach of shorter-range "point-defense" SAMs
- Close-in (CI): short-range capability, which may place the launch platform within range of very-short-range SAMs (including shoulder-fired missiles) and antiaircraft artillery

The actual standoff range numbers associated with those groupings have a tendency to change as new defense systems emerge or older ones are upgraded, but an illustration of the generally accepted bands is found in figure 4.8.

Simple bombs have very little standoff capability and are generally referred to as ballistic (i.e., nonlifting) weapons. Once released from the launch aircraft, simple bombs fall toward the surface under the influence of gravity and air drag.

Unguided rockets are another form of ballistic weapon. They may be launched from the surface or from an aircraft, but again are used without lifting surfaces. The thrust from the rocket motor accelerates the object away from the launch platform and adds velocity to the weapon, while gravity and air drag continue to affect the flight path or trajectory of the object.

Glide weapons use lifting surfaces that permit the object to fly considerably farther than a simple ballistic trajectory. In some cases the lifting surfaces are quite small, and most of the total lift is generated by the weapon body

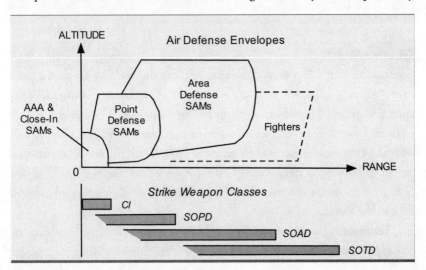

Figure 4.8 Basic Standoff Classes

itself. In other cases prominent wings are used to generate a large amount of lift. Some form of guidance is invariably used with glide weapons since their accuracy would suffer if in-flight corrections were not applied.

Longer tactical standoff capabilities are achieved by combining lifting surfaces, some form of propulsion, and guidance in the weapon. The more standoff required, the more complex the weapon. As will be seen in part 2 of this book, there are significant trade-offs associated with standoff in the design of strike weapons. For example, for the same total weapon weight, the amount of warhead carried decreases as standoff range increases, due to the larger propulsion system required to achieve that longer range.

AVAILABILITY

Strike resources available in the combat theater are highly variable. The initial inventory of weapons at the airfield or on board ship is usually based on preconflict planning assumptions about the likelihood of hostilities, the basic strike strategy during early phases of a potential conflict, and the need for routine weapons training during peacetime operations. Resupply during combat will usually reflect both desired mixes of weapons and the realities of what is actually available in resupply channels (otherwise known as the "logistics pipeline"). Theater commanders usually understand that the supply of highly sophisticated and costly long-range strike weapons is far less than the supply of simple short-range munitions. This is yet another factor that influences strike planning.

WEAPONS MATRIX

Continuing with this generic discussion, one can theoretically arrive at a multipart matrix of strike weapon possibilities when you combine warhead, guidance, and standoff variables, as illustrated in figure 4.9. The left and center columns relate to the characteristics of the target, while the right column deals with the target defenses. Due to practical considerations, not all of the potential combinations have actually been implemented, but the matrix serves as a reminder of the diversity that can be found in strike weapon inventories around the world.

There are, of course, several other ways to categorize strike weapons. In one approach weapons are grouped by their basic role or mission (i.e., target application). Such an approach results in a list of a small number of weapons

Warhead Damage Mechanism	Guidance Mode	Standoff Class
Kinetic Energy	Unguided	CI
Blast	Command	SOPD
Fragmentation	Preprogrammed	SOAD
Fire	Homing	SOTD
Other	Data Link	
Appropriate to the Target		*Sufficient to Survive Defenses*

Figure 4.9 Generic Strike Weapons Matrix

devoted exclusively to ship targets (antiship weapons), a few weapons devoted exclusively to radar targets (antiradiation missiles), and a large number of weapons that are suitable for use against multiple types of strike targets.

Yet another approach to categorizing strike weapons focuses on the level of complexity of the weapons. At the simplest level there are unguided freefall (ballistic) weapons such as basic bombs, cluster weapons, and unguided rockets. The next level of complexity would include freefall weapons with limited guidance, sufficient to reduce dispersion and cause impact at a predetermined set of coordinates or a spot designated by a laser. Then would follow glide weapons with various forms of guidance. Finally, at the most complex level would be powered guided missiles.

In point of fact, there is no single weapon categorization approach that is the "correct" method for all occasions. The author therefore elected to use standoff class as the initial screening factor, starting with the longest-range class and working inward from there, ending with the shortest-range class. This generally means that the most complex and most costly strike weapons will appear in the first standoff class, with the least complex and costly items appearing in the final standoff class.

REPRESENTATIVE U.S. STRIKE WEAPONS

What follows is a brief introduction to selected U.S. strike weapons, intended to convey the general level of equipment diversity that has been available to the U.S. Air Force, the U.S. Navy, and the U.S. Marine Corps. The weapons are grouped by standoff class. Some are marked with an asterisk, indicating that they are obsolete and no longer in service; they have been included here for historical context.

The formal military designation is used to identify each weapon, followed by its popular name or acronym. Weapons are officially designated using the following standardized nomenclature:

AGM-xx: air-launched, surface attack missile

BGM-xx: multiple launch platforms, surface attack missile

CBU-xx: cluster bomb unit; applies to dispenser weapons with submunitions

GBU-xx: guided-bomb unit; applies to guided versions of ballistic weapons

RGM-xx: ship-launched, surface attack missile

UGM-xx: submarine-launched, surface attack missile

The numbers that are found after the three letters are assigned sequentially by the Department of Defense. Thus the development of the AGM-45 missile took place before the AGM-88 missile. A letter often follows the assigned numbers to designate the sequential series of the particular item. For example, the AGM-65 missile first appeared as the AGM-65A, with later versions, some with very substantial changes, appearing as AGM-65B, AGM-65C, and so on.

Some of the listed weapons are used by all three armed services, while others are unique to a single service. Many of the short-range weapons, such as bombs and laser-guided bombs, are used by USAF, USN, and USMC aviation units, whereas ship- or submarine-launched weapons are strictly USN items.

This initial listing has been kept intentionally basic, providing no detail on the capabilities or characteristics of the individual weapons. Such information is found in appendix D. At this point it is sufficient simply to point out the array of potential weapon alternatives that warfighters consider when tasked to conduct a strike.

SOTD Class (standoff outside theater defenses; very long range)

AGM-86C C-ALCM (conventional air-launched cruise missile). A variant of the
USAF AGM-86 carrying a nonnuclear warhead; jet propelled.

BGM-109C Tomahawk TLAM-C (Tomahawk land attack missile; convention-al). A variant of the USN ship- and submarine-launched BGM-109 carrying a nonnuclear warhead; jet propelled.

SOAD Class (standoff outside area defenses; long range)

AGM-84A-D Harpoon. An air-launched AGM-84 optimized for antiship operations; jet propelled.

AGM-84E SLAM (standoff land attack missile). An AGM-84 variant modified for use against land targets; jet propelled.

AGM-84H SLAM-ER (SLAM–expanded response). A considerably modified AGM-84 optimized for land targets; jet propelled.

AGM-88 HARM (high-speed antiradiation missile) and AARGM (advanced antiradiation guided missile). Air-launched, rocket-propelled ARMs.

AGM-142 Have Nap. A USAF-only air-launched strike weapon developed in Israel; rocket propelled.

AGM-158 JASSM (Joint Air-to-Surface Standoff Missile). A stealthy air-launched strike weapon; jet propelled.

RGM-84 Harpoon. A USN-only ship-launched antiship missile.

UGM-84 Harpoon. A USN-only submarine-launched antiship missile.

SOPD Class (standoff outside point defenses; medium range)

*AGM-45 Shrike.** An air-launched ARM; rocket propelled.

*AGM-62 Walleye.** A family of air-launched glide weapons.

AGM-119 Penguin. An air-launched antiship weapon of Norwegian design; rocket propelled.

AGM-130. A USAF-only, air-launched weapon; rocket propelled.

AGM-154 JSOW (Joint Standoff Weapon). A family of air-launched glide weapons.

CI Class (close-in; short range)

*AGM-12 Bullpup.** An air-launched, command-guided weapon; rocket propelled.

AGM-65 Maverick. A family of air-launched weapons; rocket propelled.

AGM-114 Hellfire. A lightweight, air-launched weapon; rocket propelled.

*AGM-123 Skipper II.** An air-launched, rocket-propelled, laser-guided bomb.

CBU-59 APAM (antipersonnel-antimatériel). An air-launched bomblet dispenser; ballistic (i.e., no propulsion, no guidance).

CBU-87 CEM (combined-effects munition). An air-launched bomblet dispenser; ballistic.

GBU-10, GBU-12, and GBU-16. Air-launched, laser-guided bombs; no propulsion.

GBU-15. A USAF-only air-launched glide weapon.

GBU-22 and GBU-24 LLLGB (low-level laser-guided bomb). Air-launched LGBs; no propulsion.

GBU-30, GBU-31, and GBU-32 JDAM (Joint Direct Attack Munition). Air-launched, GPS-guided bombs; no propulsion.

GBU-39 SDB (small-diameter bomb). A reduced-size, air-launched, GPS-guided bomb.

M117 and M118 bombs. Obsolescent general-purpose bombs; ballistic. M is an older U.S. designation for "munition."

Mk 20 Rockeye. Older air-launched bomblet dispenser; ballistic. Mk (Mark) is a U.S. Navy designator.

Mk 81, Mk 82, Mk 83, and Mk 84 bombs. General-purpose bombs; ballistic.

The final item in this chapter is a set of side-view line drawings of selected examples of U.S. strike weapons. The weapons portrayed in figure 4.10 are all drawn to the same scale so that readers can get a sense of the relative size of these examples, which are grouped by broad generic type: ballistic weapons, guided bombs, glide weapons, antiradiation missiles (ARMs), and powered strike weapons.

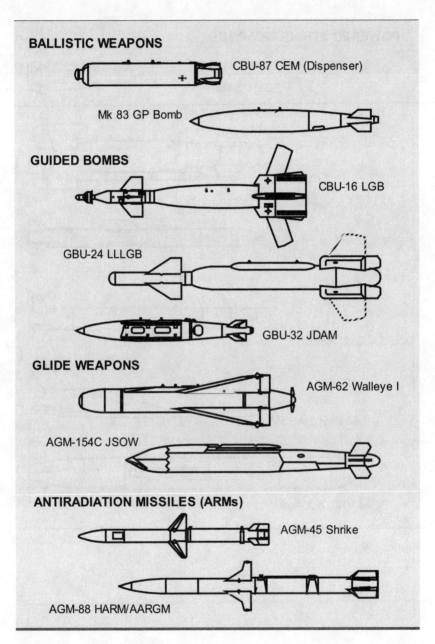

BALLISTIC WEAPONS

CBU-87 CEM (Dispenser)

Mk 83 GP Bomb

GUIDED BOMBS

CBU-16 LGB

GBU-24 LLLGB

GBU-32 JDAM

GLIDE WEAPONS

AGM-62 Walleye I

AGM-154C JSOW

ANTIRADIATION MISSILES (ARMs)

AGM-45 Shrike

AGM-88 HARM/AARGM

Figure 4.10a Example Strike Weapons

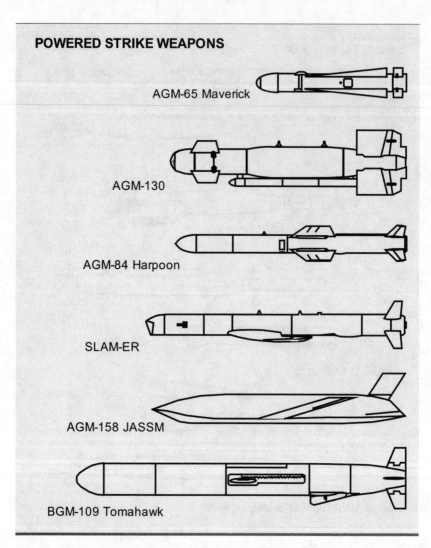

POWERED STRIKE WEAPONS

AGM-65 Maverick

AGM-130

AGM-84 Harpoon

SLAM-ER

AGM-158 JASSM

BGM-109 Tomahawk

Figure 4.10b Example Strike Weapons

FIVE

Strike Support

The launch platforms and weapons that constitute the actual strike elements are generally only one part of the overall array of forces that are involved in conducting a strike operation. There always are a number of other elements that provide important support functions. Some of those elements, such as reconnaissance satellites, airborne and surface listening posts, communications systems, and a host of logistics elements, may be far removed from the scene of the battle, and will not be addressed in this chapter. Instead, the emphasis here will be on items that directly affect the particular launch platforms and weapons involved in a specific, individual attack.

Direct Support

Certain support elements are required to properly plan and then conduct the strike. Their level of involvement or importance may vary somewhat between the types of platforms and the types of weapons involved.

PRESTRIKE TARGETING

Every strike has a target, and the attacking elements must have sufficient information about the target and its location to be able to direct the appropriate weapon to the target. Ultimately it does not matter whether the strike weapon is a ship-launched cruise missile or an air-launched bomb; the target location must be provided to the launch platform in order for the operation to begin.

More will be said about this overall issue in the next chapter, but here we simply need to understand that one essential element of direct support involves the acquisition of target location data. This process, often called "targeting," can make use of a variety of information sources. The goal is to provide the launch platform with detailed target location coordinates, along with sufficient other target characteristics and target area information to allow the attacking unit to properly plan and execute the strike.

AIRBORNE STRIKE SUPPORT

Strikes that are launched from ships or submarines generally have no additional strike support elements involved during the actual operation. The missiles launched from the sea tend to navigate their way to the target area without any accompanying support. However, strike operations using air-launched weapons often do involve other aircraft performing support functions.

Tanker aircraft are used in strike operations to refuel strike and supporting aircraft en route to or from a target area. Airborne early warning (AEW) radar aircraft are also commonly employed to maintain surveillance of the airspace in the region and help coordinate air operations.

If there are substantial defenses to be faced en route to the weapon launch point(s), there may be a significant number of other aircraft involved in various forms of protection for the strike aircraft (the "strikers"). Considerably more will be said about this subject in chapter 7, but a brief overview will be found here just to put the forces in context. Fighters may precede or accompany the strikers to engage enemy fighters. Other aircraft may employ electronic countermeasures (e.g., jamming) to screen the attacking force from enemy radars; these electronic-warfare (EW) aircraft may accompany the strikers or may be offset to establish the desired jamming geometry. Finally, some aircraft may provide defense suppression support by directly attacking enemy air defenses (e.g., SAM sites) using antiradiation missiles (ARMs). In current military parlance, this mission is known as suppression of enemy air defenses (SEAD).

Overall, it is not unusual to find that the strike support elements far outnumber the strikers in a given operation. This would be particularly true in situations where the defenses are formidable in both capability and depth.

POST-STRIKE EVALUATION

After the strike has been accomplished, there remains the matter of evaluating its effects. Some basic questions need to be addressed before the operation is considered complete: Did the strike weapons reach the assigned target? Was the target damaged? To what degree? Were nontarget objects damaged? Will a follow-up strike be needed?

This damage assessment portion of the operation generally begins with surveillance of the target after the smoke and dust clear. It may involve the same resources used in prestrike targeting, but it often poses a more difficult challenge than simply identifying a target location and basic target characteristics. It is sometimes very difficult to determine accurately the level of damage that may have been inflicted inside a large, physically robust target. In some cases the level of activity around the target may have to be observed for a period of time to infer what might have happened inside.

Sustaining Support and Resource Management

Except for the occasional punitive action, strike operations seldom occur as single events. They are usually associated with some ongoing campaign that may include a great many such operations. As a result, all of the direct and indirect support operations must be considered in the context of a continuing series of events, often overlapping, sometimes conflicting, and generally at a high tempo. This will become more apparent in the discussion of strike planning in chapter 9.

While strike operations are conducted in a continuing cycle of planning, execution, and assessment, an important parallel operation on the logistics side involves resupply of war matériel. Weapons, for example, must be ordered from stockpiles or sources outside the theater of operations, must be transported to the theater, distributed to combat units, and prepared for use. There are limits to available numbers of replacement weapons, transports, and the like, all leading to constraints on what may be available to any given combat unit at any particular time.

Field commanders are therefore faced with very real challenges in balancing needs with available resources. When the preferred weapon is not immediately available, they may have to compromise and use something less desirable, which may require additional strike support, and so on. Real-time management of strike resources is an art form that taxes the skills, patience, and creativity of combat units.

SIX

Finding the Target

In the not-so-distant past, strike warfare was dominated by heavy shore bombardment from ships and saturation, or "carpet," bombing from large aircraft. Massive amounts of firepower were heaped onto the target area, basically attempting to level the place and thereby ensure that the real target(s) would be destroyed in the process. This kind of practice involved bringing the launch platforms into close range and then using their bombsights or gunfire directors to send the munitions on their way.

As ballistics and atmospheric effects became better understood, it became practical to single out the specific target and launch a smaller number of weapons against that object. The inherent dispersion of those unguided strike weapons still caused multiple weapons to be used to improve the probability that the target would be destroyed, but the amount of unnecessary damage to adjacent areas was reduced considerably.

When guided weapons were introduced, it finally became realistic to contemplate launching a single weapon to destroy a single target. With time, the phrases "precision strike" and "surgical strike" became not only part of the popular vernacular but also valid descriptions of the capabilities of some of those guided strike weapons.

This evolving economy of force comes with a price. It now becomes necessary to add complexity and cost to the weapon, as guidance and control functions are introduced. If weapon standoff is needed, even more elements must be added to provide lift and propulsion.

Yet another impact of increased weapon accuracy or standoff is a need for improved target information. Expressed in aviator terms, "You can't deliver a weapon to the target if you don't know where to send it." Even the U.S. Postal Service needs a valid address to properly deliver a package.

Essential Target Data

Just as in the real-estate world, the essential ingredient is "location, location, location." Determining exactly where the target is situated is a vital step in the operational process. In some cases this action is taken by the launch platform itself when it uses its onboard sensors to find the target and determine its position relative to the platform. For example, an attack of a ship target may require that the launch platform employ its search radar to scan the ocean, find the target, and use that radar data to obtain range and bearing from the platform to the target. When the target location is described in terms of distance and angle from the sensor, it is said to be expressed in "relative coordinates."

In other cases target location information may come from sources other than the launch platform. Under these circumstances the location data must be transferred from its original source to the unit assigned to carry out the attack. Such information is usually expressed in "absolute coordinates," meaning that the location is specified in some compatible map reference grid system such as latitude and longitude.

The degree of accuracy required in target location data varies with the type of strike weapon. Those that use guidance modes that can actually search for the target within a given area of water or land can usually tolerate moderate inaccuracies in target location data. On the other hand, those weapons that simply navigate to a specified location must be provided with very accurate target location data if the attack is to be successful.

In addition to location, there are some other items of target information that are helpful to the attacking forces. As will be seen in chapter 9, "Strike Planning," it is important to know something about the size, configuration, construction, and functional elements of the target. A small, lightly constructed target may be dealt with using a single small weapon. A physically large target, such as a manufacturing plant, may warrant multiple weapons impacting at different locations in the building, or perhaps a single weapon impacting a particularly critical functional element within the building.

Large structural targets introduce another aspect of the question of location. When the target location is specified, it is important to know exactly where on the target that location applies. Is it the center of the target, a particular corner, or some other reference point? For example, when you are attacking a sixty-foot by four-hundred-foot bridge using a guided weapon that promises delivery accuracy to within a few feet, it would be helpful to know exactly where the location coordinates are centered.

Targeting Resources

Many different resources are used to obtain such "targeting" information. Perhaps the most exotic of these are reconnaissance satellites, sometimes called "overhead assets" or "national assets." The easiest way to envision these sensors is to picture them as a Hubble Space Telescope with the telescope pointed toward Earth, producing images that make Google Earth seem archaic. While visual images are one important output, other important products include infrared images, radar scans, and electronic intercepts.

Surveillance aircraft and UAVs are shorter-range airborne sources of targeting data. They, too, can acquire visual, infrared, radar, and electronic data within the capability of their sensors. Surface-based radars and electronic-intercept sites, both ashore and afloat, add to the body of information that can be applied to the targeting issue.

All of the resources noted above are categorized as "off-board" elements, meaning that they are separate from the launch platform. Information collected by these off-board resources must be forwarded to strike planners. Because of the security classification involved in many of these surveillance and reconnaissance operations, not all of the material collected may be forwarded to the ultimate user. Only the data absolutely essential for strike mission planning and execution may actually get passed along.

"Onboard" targeting resources are those found on the launch platform itself. These may include visual, infrared, radar, or passive electronic ("listening") sensors on either ships or aircraft carrying the strike weapons. In the case of aircraft platforms, these onboard sensors come into play after a strike mission has been planned and is already under way.

Targeting Issues

If much of the above sounds like an intelligence-gathering operation, it is. Very large sums of money have been expended to establish and maintain a robust surveillance and reconnaissance program to obtain as much information as practical about potential adversaries. In most cases the collection of such information is done in a covert manner to avoid signaling our interest in particular objects or areas. Both the sensitivity of the subject matter and the sensitivity of the equipment and processes lead to very high security classifications, which in turn severely limits the availability of the complete library of collected material.

A recurring concern with targeting materials involves the difference between resolution and location accuracy. For example, intelligence sources may provide a very clear and detailed high-resolution photograph of the intended target, but unless accurate location data is also available, there may be insufficient information to conduct a strike using long-range weapons. That is to say, the photograph may present marvelous detail about the exterior of the target, but we may not know the target coordinates to the needed level of accuracy. In a few instances the reverse may be true. We may have very accurate location data on a vague or unimaged object, such as a possible entrance to a suspected underground bunker. The ideal situation occurs when both high-resolution imagery and a precise location reference is available for the target.

Some targets are relatively easy to find and characterize. A large building of known purpose situated in an uncluttered, open area would be one such example. Other targets are much more difficult. Items that are small, mobile, in the midst of clutter, camouflaged, and so on pose greater challenges. Targets that are buried, such as command bunkers or other critical facilities, are troublesome also, even though they are stationary.

When a region has been identified as a potential conflict zone, surveillance and reconnaissance resources may be tasked to monitor the area and collect data on specified kinds of facilities. As information is collected from various sources, including human intelligence sources on the ground, analysts attempt to catalog and summarize the material in formats usable by both decision makers and military forces. Over time a candidate target list usually emerges, drawn from what military operators call the "order of battle" and expanded to include other important (nonmilitary) elements such as industrial, energy production, and infrastructure targets.

Efforts to collect, analyze, and summarize such information are quite extensive and time consuming. But the investment can be extremely valuable if a conflict should occur in a region that has been studied. The earlier preparations provide the tactical commanders an advantage by immediately making available a considerable amount of information on potential targets. This is particularly true of objects that are judged to be "high-value" targets and thus likely to be attacked during the early phases of a conflict.

Finally, some targets are simply unknown to planners before a conflict, and many targets are sufficiently mobile that they cannot be specifically located ahead of time. Attacking such objects typically involves "real-time targeting" efforts, where onboard or off-board sensors are employed to obtain the necessary location data in the midst of ongoing military operations.

SEVEN

Defeating the Defenses

The principal goal of strike warfare practitioners is to destroy assigned targets without suffering losses of launch platforms or support elements. In warrior speak: "Kill the target and get home safely." Achieving that goal is not very difficult when defenses are light, but the presence of modern air defense systems, manned by competent and dedicated opponents, creates a much more challenging situation. The worldwide proliferation of very capable air defense systems means that deadly counterfire can be introduced in any theater of operations, at any time. It is this "high-threat" environment that sets the bar for evaluating the survivability of both launch platforms and strike weapons. If the platform is turned back or destroyed by the defenses before weapon launch, the mission is a failure. Similarly, if the strike weapon is destroyed by the defenses before reaching the target, the mission is also a failure. Neither case is desirable, but the impact of losing platforms and crews is clearly the more critical.

Fundamental Strategy

At the most basic level, the strategy behind survivability enhancement is to limit exposure and risk. If the air defenses are not able to detect the attackers, the attackers win. If the air defenses cannot reach the attackers, the attackers win. If the air defenses are not able to successfully engage the attackers, the attackers win. And so on.

Knowledge of the air defenses is critical to defeating them. The more that the attacker knows about air defense locations, performance capabilities, resupply availability, operational doctrine, functional peculiarities, and vulnerabilities, the easier it becomes to successfully deal with those defenses. Intelligence operations focusing on air defense systems are therefore a high priority, and surveillance resources are often tasked to keep track of particularly troublesome air defense elements such as long-range surface-to-air missile (SAM) sites.

If the attacking force cannot avoid the air defenses, then it attempts to disrupt, disable, or destroy them. This can involve several techniques, described below, and is a dedicated effort to reduce the effectiveness of the air defenses along the route of flight of the attacking force. This kind of strike support is not typically used to protect long-range strike weapons such as cruise missiles, but is generally reserved only for the protection of the launch platforms.

Steps taken to defeat or negate defenses are often grouped under the general term "tactics." Tactics are formulated based on the nature of the defensive threat and the capabilities of the attacking force. For example, if the threat is substantial and the attacking aircraft are relatively easy to detect, then it is likely that a large amount of support will be employed to allow the aircraft to reach their launch points and return safely. On the other hand, a stealthy attack aircraft would need far less support when facing the same threat.

Overall, there are four general categories of protective techniques employed by strike forces. They are seldom used in isolation, and more often are combined and integrated into the strike plan for the particular mission.

Avoidance

The first technique is simple avoidance. If you know where the defenses are located and have the flexibility to go around them, do so. The route-planning portion of the strike plan is often driven by this basic technique. Enemy radar and SAM site locations are plotted on route maps, and their coverage envelopes are added to give the strike force an indication of the high-threat zones. Routes can then be laid out that avoid or minimize exposure to the defenses, and where exposure cannot be avoided, other measures can be planned.

Stealth enters the strike-planning process at this point. Contrary to popular belief, stealth aircraft are not truly invisible to radar, but they appear so much smaller than conventional aircraft that the effective range of the radar is reduced to a small fraction of its normal level. Thus the radars associated with a defense system may be degraded to the point of being virtually ineffective against a stealth aircraft. The following figure may help visualize the effect.

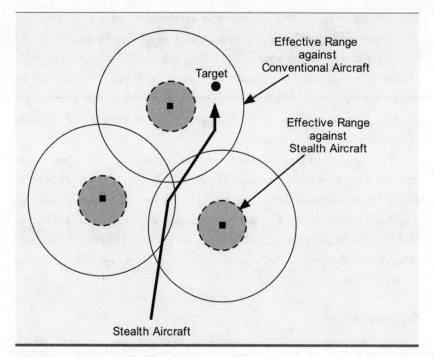

Figure 7.1 Effective Detection Ranges

Figure 7.1 indicates that the several radars have overlapping coverage (solid circles) when dealing with conventional (large-signature) aircraft. But coverage against a notional stealth aircraft shrinks markedly (dashed circles), meaning that large areas are now unprotected. Stealth basically opens up holes in the combined air defense envelope, allowing a stealth aircraft to safely operate in areas where conventional aircraft would be in mortal danger. Even if briefly detected, the exposure time for a stealth aircraft may be so short that the air defenses cannot respond effectively. The same would be true for any stealthy strike weapons; these should be able to fly into extremely high-threat environments with greatly increased survivability.

Deception

A second basic technique is to introduce confusion and deception into the enemy air defense picture. This involves the creation of distractions, to cause the defender to look away from the actual strike route and focus attention somewhere else, at least for a critical period of time. The effect is similar to that of a stage magician or illusionist, who draws the attention of the audience to a phony activity while the important action occurs elsewhere.

In strike warfare the attention-grabbing activity may consist of a group of aircraft maneuvering as if preparing to attack from an area away from the primary strike route. Or, instead of a feint using actual aircraft, it may involve the launch of airborne decoys to simulate a group of strike aircraft heading inbound into defended airspace.

Such deceptions have been employed in conflicts, and have been quite effective on at least some occasions. However, they are not generally popular with military operators for two reasons: (1) They require the participation of additional aircraft that may be in high demand, and the expenditure of decoys that aren't freely available; and (2) the aviators involved usually aren't terribly pleased with the assignment because, as one put it, "Nobody likes shooting blanks in a combat zone."

Suppression

A more direct response to air defenses is found in the third technique, which actually encompasses both electronic and high-explosive elements. The term "electronic warfare" (EW) has historically been associated with the use of dedicated equipment to jam enemy radars and other electronic devices. As radars have become more sophisticated, and thus able to reject simple electronic noise, the jammers have likewise become more sophisticated.

Jamming aircraft may be part of the strike support force assigned to protect the launch platforms. The jamming is usually intended to screen the attackers from view by specific defense radars. However, jamming is not a panacea, for several reasons. First, it is largely directional, meaning that jamming that is aimed at one radar only affects part of that radar's overall view of the sky and does not necessarily provide much protection against other radars that are offset from the targeted radar. Second, jamming attracts attention. Once jamming has begun, it is clear to the defenders that something is afoot, and their air defenses become very focused on the area around the jamming source. Third, modern jamming generally requires dedicated aircraft and crews whose sole function is electronic warfare. Such assets are typically few in number and may have to be allocated on a priority basis. Finally, jamming does not destroy the targeted radar; it simply degrades its performance for a short period of time. This means that the attacking force may have to deal with that particular radar again on a subsequent mission.

A more destructive defense suppression technique is the use of antiradiation missiles (ARMs) to attack the radar transmitting antennas of the air defense sites. The best-known examples of these are the AGM-45 Shrike missile, first used in North Vietnam, and the subsequent AGM-88 HARM (high-speed antiradiation missile). When the missile impacts or passes by the radar antenna, the high-explosive warhead detonates and sends a spray of fragments into the area, destroying equipment in the vicinity. The original practice was to employ ARMs from dedicated defense suppression aircraft whose mission was to "listen" for the telltale radar emissions from SAM sites and then launch weapons against those radars. More recent practices place ARMs on some jamming aircraft or on some strike aircraft, depending on the circumstances of the particular strike mission.

Another concept for disabling the defense radar site would involve a different kind of warhead. Such a weapon would still home in on the radar transmissions, but instead of blasting the antenna and nearby items with fragments, there would be an intense burst of electronic energy, intended to overload and destroy the electronic circuitry. This "high-power microwave" (HPM) warhead approach leaves most mechanical components, such as the antenna, undamaged, but would cause a functional failure of the radar electronics.

While radars remain a vital part of air defense sites, there are other elements that are also worthy of attack. Command and control of the site may be in a separate vehicle or building; if this element is destroyed, the site becomes inoperative. The SAMs themselves are worthwhile targets, either those on their launchers or those in ready storage. Destruction of these nonradar elements generally requires use of strike weapons.

Self-Defense

As a last-ditch technique to enhance survivability, the strike aircraft generally do have several onboard protective measures available to them. Small bundles of radar chaff may be released from the aircraft to confuse radar-guided antiaircraft missiles. Chaff is radar-reflective material that disperses in the air, creating a substantial return when viewed by radars or radar-guided missiles. Similarly, small pyrotechnic flares producing intense infrared radiation may be ejected to draw off infrared-guided missiles. Both types of devices are ejected from dispenser units buried within the aircraft.

In more recent years it has become common practice to equip strike aircraft with towed decoys. These are small objects that are reeled out behind the aircraft when it approaches defended airspace. The decoys mimic the aircraft signature, causing air defense missiles to aim at a phantom aircraft a safe distance behind the actual aircraft.

Active protection using an onboard laser is a new technology that may eventually become commonplace. The concept here is to focus laser energy on an infrared-guided antiaircraft missile that is homing on the aircraft, thereby causing the missile seeker to quit homing in on the aircraft (known as "break lock") and miss.

EIGHT

Rules of Engagement

Despite what may seem to be appearances to the contrary, most state-sponsored military operations are conducted in accordance with international law, conventions, and protocols. Efforts to establish limits and conditions for armed conflict began in medieval Europe and continue to this day. At the present time the most well-known "laws of war" are found in the several Geneva Conventions, the Hague Conventions, and the more-recent United Nations protocols. These various agreements focus largely on the humane treatment of wounded, prisoners, and noncombatant civilians. Some forms of combat or weaponry are prohibited, and there is a general requirement that combatants be visually identifiable by the wearing of distinctive uniforms or markings. If nations at war are signatories to such agreements, they are both bound to abide by the terms and are protected by them. However, what is not widely understood outside of the military community is the fact that parties that are not signatories (e.g., terrorists, insurgents, or nonsignatory nations) are neither forced to comply with nor protected by the terms of these conventions.

These broad laws of war are seldom a direct issue in the conduct of strike warfare. However, they may affect the establishment of what are known as the rules of engagement (ROEs) for a particular conflict. These self-imposed stipulations form a critical set of criteria that are applied during the planning and conduct of strikes, and they can have a powerful influence on mission alternatives and its ultimate success or failure.

Basic Definition

Military ROEs are written directives that specify when, where, and how force may be used in a particular conflict or portion of a conflict. They are issued by the senior leadership in a conflict, and represent top-level instructions to subordinate commands. As such, they may be established for operational, economic, humanitarian, or political reasons.

Rules of engagement may be very narrow in applicability, such as a specific locale, time frame, or mission, or they may be quite broad, encompassing an entire theater of operations for the duration of the conflict. They basically form a legal framework within which armed forces may operate. Violation of rules of engagement can lead to disciplinary action, including court-martial.

Examples

In any given conflict, it would not be unusual to find an overall set of rules of engagement for the general conduct of operations, supplemented by subordinate ROEs for specific regions or actions. In each case the rules may change with time, reflecting changes in conditions, objectives, or policy. With that in mind, it is important that individual combat units be kept advised of the latest ROEs so that inadvertent violations do not occur.

At an overall level, the rules of engagement may provide broad guidance on basic issues, such as the following hypothetical examples:

- Combat units shall always have the right to use deadly force in self-defense, regardless of who is attacking the unit.
- Operations shall be confined within the internationally recognized borders of [name of country] and adjacent waters; no combat unit shall enter the territory of adjacent countries.
- Nuclear, chemical, or biological weapons shall not be employed.
- Civilian casualties are to be minimized.
- Damage to religious sites is to be minimized.
- Beyond-visual-range (BVR) air engagements, both surface to air and air to air, are authorized provided that hostile identification of the target has been established.

At the subordinate level, there may be a set of generalized rules of engagement dealing with ground operations, addressing such things as response to threats in civilian areas, treatment of noncombatants, seizure of property, and so on. Continuing on, a more-focused set of supplementary ROEs dealing with a particular strike mission might include items such as the following hypothetical examples:

- Collateral damage to immediately adjacent buildings is acceptable if necessary to achieve mission objectives.
- Minimize collateral damage to nearby civilian areas.
- No weapons may be released until final clearance is received from the local forward air controller (FAC).

Impact on Operations

Rules of engagement are intended to place certain constraints on military operations. Some of those constraints are widely accepted as matters of common decency (e.g., minimizing civilian casualties) or pragmatic sensibility (e.g., no incursions into neutral countries). In other cases the rules may represent sometimes-controversial policy decisions. One of the most well-known examples of the latter comes from the Vietnam era, when a conscious decision was made on the part of the upper civilian leadership of the United States to not bomb specific targets in North Vietnam. Only those targets on the approved target list could be attacked, even in the face of evidence that more-lucrative targets might exist on the prohibited list.

During the Korean conflict of the 1950s, United Nations aircraft were prohibited from pursuing Communist MiGs into Chinese airspace when the enemy fighters returned to their airfields located there. This gave the Communist forces a notable advantage, allowing them a basing and airspace sanctuary from which to safely operate.

In yet another example, there have been occasions when U.S. aircraft have been forbidden by ROEs to engage other aircraft unless positive visual identification has been made. While hostile identification can be accomplished at standoff distances by several means, positive visual identification can only be achieved by close inspection. This means that U.S. aircraft must approach enemy aircraft at extremely close range before they can fire. In effect the ROEs negate the value of medium- or long-range air-to-air missiles, and forces American pilots to use only guns or short-range missiles. Enemy aircraft, on the other hand, are generally under no such restriction and can engage at will, placing American forces at a distinct disadvantage.

Awareness of ROEs

As noted earlier, it is important that friendly forces be kept informed of the current rules of engagement. Extending that knowledge to the enemy or to noncombatants is quite another matter, however. Selective release of information is the more-likely course of action.

It may be advantageous to make known certain basic ROEs, including those dealing with respect for borders, treatment of civilians, and so on, as a matter of building confidence about American intentions in the region. There may also be sound rationale for making known certain operational ROEs that identify what U.S. forces will or will not do under particular circumstances (e.g., a return-fire policy even if the fire comes from a religious site). This may serve as fair warning to those who might otherwise try to exploit a normally sensitive issue.

Knowledge of other ROEs may best be restricted to friendly forces only, to avoid providing the enemy too much information about tactical plans. ROEs that are related to specific operations are generally in this restricted category, at least until well after the operation is over.

NINE

Strike Planning

Some adventure novels or action motion pictures portray the strike preparation process as a brief, rather casual affair that occurs with minimum impact on the plot or storyline. The squadron leader says, "We gotta get rid of that bridge. You guys load up some bombs and go take care of it." And in the next scene, the attacking force is heading for the target.

That kind of fictional depiction unfortunately leaves out some very important realities, beginning with the matter of who makes the decision to attack a particular target. It also tends to focus on a single strike, when in fact there are most likely multiple strikes under way at any given point in time, with a number of others in preparation. Fictional accounts also leave out much of the depth of preparation necessary for a successful strike. Some of those glossed-over details are briefly touched upon in this chapter.

The term "strike planning" actually encompasses several levels of preparation for attacks on surface targets. It begins in the upper levels of the command hierarchy where broad decisions are made regarding overall military objectives and policies, which lead to basic guidance on rules of engagement and basic classes of targets to be attacked. At an intermediate level, that broad guidance gets translated into prioritized strike objectives within the area of responsibility. This leads to the generation of specific strike assignments, which are coordinated at the intermediate level before being passed to the combat units that will actually conduct the individual strikes. Those individual units (e.g., ships, squadrons) then generally plan the details of the strike, within the constraints imposed on them by higher authority.

This chapter addresses the planning process in two parts, beginning with the upper- and intermediate-level issues and then dealing with the planning for a specific strike by the combat unit tasked to conduct the strike.

Strike Tasking and Coordination

The top-level objectives, strategy, and rules of engagement are the foundation for subsequent steps in planning and conducting combat operations. These form the "big-picture" guidance for what lower-echelon combat units are expected to accomplish during the conflict. Such guidance can be considered a framework that is then built upon by the responsible units farther down the chain of command.

An example of a top-level objective might be to "neutralize enemy surface-to-surface missile capabilities." At the intermediate level, this might be translated into a need to conduct a series of strikes against various known or suspected SSM sites or staging areas. Ongoing operations and available resources would be reviewed and coordinated before assignments are made to designated combat units to actually execute strikes against assigned targets.

The intermediate-level activity is dominated by a focus on resource management. There are limits to available people, platforms, weapons, and so forth, which must be balanced against combat needs in a dynamic, changing situation within the theater of operations. Strike warfare is generally just one of several competing missions, which means that difficult decisions are often needed to maintain a successful overall campaign. As one senior staff member once described it, "The process resembles medical triage at the scene of a disaster, where everything gets prioritized and attention is given to the most pressing problems . . . but the problems keep changing."

Strike warfare is often broken down into several subtypes at the intermediate level. These have historically included:

- Deep strike or basic strike: attack of major targets on land
- Antisurface warfare (ASuW): ship attack
- Interdiction: attack of mobile targets or supply targets on land
- Suppression of enemy air defenses (SEAD): attack of air defense sites
- Close air support (CAS): on-call attack of battlefield targets, often in close proximity to friendly forces

Essential support of airborne strikes comes from additional resource categories that also must be coordinated. These include:

- Aerial refueling: airborne tankers
- Antiair warfare (AAW): fighters and airborne early warning (AEW) assets to cope with enemy fighters
- Combat search and rescue (CSAR): units on standby to extract downed aircrew members

One of the keys to successful overall operations is a timely flow of useful intelligence information to the planners and decision makers. Just as important is accurate and up-to-date information on the status of friendly forces. In both cases the information changes frequently, sometimes dramatically, which can cause priorities and assignments to be revised. Faulty information or gaps in information are another reality, leading to the phrase "the fog of war," which aptly summarizes the uncertainty that frequently plagues combat operations.

Nevertheless, missions do have to be defined and assignments made. The intermediate level is where specific strike targets are selected for attack. This might mean a specific bridge, rail yard, communications facility, or the like, or it might mean an attack on a moving force of armor or group of ships. The best available information on the target location is specified, along with any special rules of engagement (over and above the generalized ROEs in effect in the area). Supplementary information on the target itself, the target area, and known nearby defenses is collected and added to the package. Finally, two very important items are added. First is the required "time over target" (TOT) or a target attack time "window"; this reflects a basic level of combat coordination, or "deconfliction," to minimize inadvertent mixing of more than one attack force in the same area at the same time. Second is the required level of damage to the target, often expressed as the "probability of kill" (Pk) or "probability of damage" (Pd).

This package of material forms a strike assignment that is then provided to the combat unit charged with the responsibility of executing the strike. When airborne units are involved, this is generally called an "air tasking order" (ATO), and the responding units are said to have been "tasked" to conduct the particular strike. But the strike plan is not complete at this point, as will be seen in the next section.

In practice the intermediate-level strike-planning process is under way twenty-four hours a day, seven days a week, with multiple missions being addressed simultaneously. It is an overlapping process in which a number of small teams work in parallel. As each team completes one mission assignment, it is handed another. The starting times and completion times for mission assignments do not coincide between the small teams, leading to a virtually continuous flow of material into and out of the planning facility. There is no such thing as downtime.

Individual Strike Planning

The combat unit assigned to execute the strike hasn't just been sitting around, waiting for something to do. It typically has been busy handling other tasks, and it will surely be busy with other actions after the current assignment is addressed. The new tasking orders just received from the intermediate level are simply fit into the continuing operational cycle.

Depending on current operational doctrine within the combat force, the specific strike plan may be prepared by the executing unit itself, or it may be prepared by an associated group of planning specialists. In either event the individual strike plan begins with the tasking order, which defines the following:

- The target (and perhaps specific aimpoints within a target)
- The required level of damage or kill
- The timing of the strike
- Basic conditions and constraints

The information package that accompanied the tasking is then expanded by the addition of whatever supplementary information may be available at the combat-unit level regarding such factors as:

- Refined target data (size, configuration, orientation, construction, etc.)
- Target area conditions (terrain, foliage, structures, clutter, etc.)
- Air defenses (target area and en route)
- Environmental conditions (forecast weather, lighting, smoke/haze, etc.)
- Navigation data for airborne units (route-planning information)

After considering the target and basic defense conditions, a preliminary attack concept is formulated that reflects the strike resources available at the combat-unit level. The selection of the proposed strike weapon(s) is a process often called "weaponeering," which matches the capabilities of the attacking weapons against the needs of the mission. Planners have access to a considerable database of weapon effects against various targets to aid in this process. The general philosophy that is followed is to use the least complicated weapon that will suffice for the assignment, an approach that warfighters summarize with the statement "Never use a silver bullet when a lead bullet will do."

A critical element in the preliminary attack concept is the position of the launch platform that will enable a successful launch of the strike weapon(s). A common approach to this issue is to start at the target and work outward, considering alternative approach paths to identify one or more preferred routes and associated launch points.

Ships or submarines launching very-long-range cruise missiles may not have to move to a new location to reach a suitable launch point. However, when weapons are launched from aircraft platforms, it is commonplace for the aircraft to have to fly some distance from their land or sea base before reaching the launch point. It is also likely that some of that airborne route will be in defended airspace, requiring careful preparation for defense avoidance and suppression. Aircraft fuel management and aerial-refueling issues must also be addressed.

The tactics employed in strikes vary with time and circumstances. Tactics tend to be driven by the defenses in place in the particular area (which may change with time) and reflect the capabilities of the attacking platforms and weapons. At an overall level, planners are mindful of basic combat fundamentals that have stood the test of time, which are best summarized in what are called the "Principles of Warfare." The exact wording may vary slightly from one source to the next, but the key elements are found in the following nine basic principles:

1. *Objective.* Direct every military operation toward a clearly defined, decisive, and attainable objective.
2. *Mass.* Concentrate combat power at the decisive time and place.
3. *Maneuver.* Place the enemy in a position of disadvantage through the feasible application of combat power.

4. *Offensive*. Seize, retain, and exploit the initiative.

5. *Economy of force*. Employ all combat power available in the most effective way possible; allocate minimum essential combat power to secondary efforts.

6. *Unity of command*. Ensure unity of effort for every objective under one responsible leader.

7. *Simplicity*. Avoid unnecessary complexity in preparing, planning, and conducting military operations.

8. *Surprise*. Strike the enemy at a time or place or in a manner for which he is unprepared.

9. *Security*. Never permit the enemy to acquire unexpected advantage; protecting the force increases combat power.

The preliminary strike plan is reviewed with the cognizant combat-unit commander once it has taken shape. If the commander concurs with the approach, the details are added and the plan is finalized. When completed, the strike plan includes both the operational aspects of the attack and the supporting elements, including identification of all necessary resources and external coordination.

Example Strike Plan

No two strike plans are exactly alike, even against similar targets under roughly similar conditions. There may be similarities in basic elements, but the details tend to differ as they reflect day-to-day changes in the combat situation. A hypothetical and simplified strike plan is illustrated below just as an arbitrary example. In this case it is assumed that a strike squadron on board a Navy aircraft carrier has been assigned the task of destroying a small electrical power generating station along the coast. The target is situated along a small inlet and is under the protective umbrella of a nearby short-range SAM site. There is also the possibility of enemy fighters in the area.

In response to this tasking, a strike plan has been generated that makes use of two moderate standoff weapons that can be launched from just inside the effective range of the SAM site. To counteract this threat, jamming will be used down the axis of attack to obscure the enemy's view of the inbound strike force. Concern over possible enemy fighters also suggests that friendly fighters be included in the strike package, to sweep the area ahead of the strikers and be in a position to intercept and engage any enemy aircraft that might appear.

The resulting plan of attack is illustrated in figure 9.1. The force departs the aircraft carrier at the location marked "A" and proceeds in a southwesterly direction. There is an airborne early warning (AEW) aircraft with nearby combat air patrol (CAP) fighters maintaining surveillance of the airspace in the racetrack pattern marked "B." The outbound strike force has fighters (VF) in the lead, followed by a jammer aircraft (VQ) and the strikers (VA).

Numbers are used in the figure to indicate the location of the aircraft at various points in time. At point "1" the fighters are considerably ahead of the other aircraft. At that time the fighters turn west, the strikers turn toward the target, and the jammer turns east. The jammer begins jamming at this point, masking the movement of the strikers as they head toward the weapon launch point.

The fighters maintain their protective positioning while the strikers continue inbound, screened by the jammer as it follows a racetrack pattern to keep the jamming energy focused on the SAM site. When the strikers reach the launch point (point "3"), the weapons are released and the strike aircraft reverse their course, staying within the area subject to jamming. About the time that the weapons are arriving at the target (point "4"), the jamming is terminated and everybody heads back toward the current position of the aircraft carrier.

This example illustrates a fairly simple strike, involving a single-axis attack, no antiradiation missile support, and no need for aerial refueling. Even so, it should be apparent that most strikes involve multiple parties engaged in a closely coordinated action that has been described by some as "combat choreography."

Taken in a larger context, there is a circular flow of activity associated with each strike. It begins with a target being selected. The planning phase includes first the intermediate-level effort, the tasking, and the combat-unit planning. This is followed by strike execution, described in the next chapter. Then comes strike assessment, to determine the results of the strike. Based on that assessment, a decision is made either to consider the mission complete

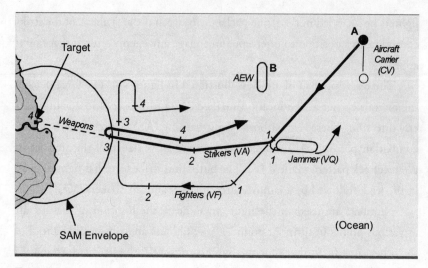

Figure 9.1 Example Strike Plan

(i.e., the target has been sufficiently damaged or destroyed) or to prepare for a reattack (i.e., insufficient target damage). If the decision is to attack the target again, it is prudent practice to avoid an exact repeat of the first attack; the enemy may have been caught unawares the first time, but the element of surprise is now gone.

TEN

Strike Execution

It might seem to some that once a thorough strike plan had been prepared, indicating a relatively safe set of conditions for the attacking forces, the actual strike should almost be an anticlimax. Such is seldom the case. Something unexpected often occurs: equipment fails, the weather changes, an unknown defense appears, and so on. It is a truism that "in combat, plans only work until the first contact with the enemy."

Nevertheless, the strike plan is the key guidance for the execution of the attack. It will be followed unless and until there is some compelling reason to alter it during the course of the mission. For example, the approach of enemy fighters might cause an inbound strike force to modify its course to avoid contact, which in turn might revise the weapon launch point. A certain degree of flexibility and real-time initiative should therefore be available to the attackers to cope with the unexpected.

There is a general sequence of steps that occur during the execution of a strike. The sequence for a surface launch attack differs from the sequence for an air launch attack, and, indeed, there are differences between types of platforms. What follows is an overview of the major steps associated with the two basic categories of platforms: surface and air.

Surface Launch Missions

The launch of a standoff strike weapon from a ship or submarine begins with the selection of the specific weapon to be used. If that weapon is in storage it will need to be brought out for preparation. However, the majority of such current weapons may already be stored in a ready condition in a launcher of some type.

The selected weapon must be checked out and prepared for launch. This may involve running what is called a built-in test (BIT) that basically checks the status, or "health," of the weapon. If the test indicates a fully functional weapon (a "GO" condition), the sequence proceeds. If faults or problems are detected by the BIT (a "NO-GO" condition), the weapon is bypassed and another one is selected for the mission.

The next step is the transfer of mission data to the weapon. This varies from system to system, but includes any route and target information that may required to navigate to the target area, find the target, and home in on it. Mission data may involve cruise altitudes, GPS waypoints, target characteristics, guidance modes, and so on.

With verification that mission data has been successfully loaded into the weapon, everything is ready for launch. When the platform reaches the designated launch point, any final prelaunch steps are taken to "initialize" the weapon, and it is launched.

A rocket booster is used to accelerate the weapon away from the launcher. In the case of a submarine launch, a buoyant capsule may be used to bring the weapon to the surface, where the rocket motor is then fired. Other types of sub-launched missiles are launched "bare" and either propel themselves using the booster motor or rise to the surface using the momentum of the ejection and then ignite the booster. In either case they rapidly accelerate to flight speed after leaving the water.

The boost propulsion phase is relatively brief and is intended to get the weapon stabilized at a speed compatible with the cruise propulsion system. In most cases the spent booster motor is detached and falls away from the weapon while the cruise propulsion (rocket or jet engine) is started. The remainder of the flight is powered by the cruise motor, and is often characterized by an extended period at a relatively constant airspeed.

The weapon navigates toward the target based on the mission data instructions it received prior to launch. This "midcourse" phase is generally accomplished without any human intervention, although it would be possible for the weapon to accept revised mission data while en route if the weapon were so designed.

Ultimately the weapon arrives in the target area. Depending on its guidance mode, it may activate some form of seeker that is intended to search for, acquire, and home in on the target, or it may simply guide to a set of target coordinates that were part of the mission data. In both of these cases the final phase of flight (called the "terminal phase") is autonomous, meaning without human intervention.

There are also methods available to permit data transfer in the terminal phase and even for human intervention if desired. These require the use of a data link between the weapon and a control or observation site located some distance away. When very long distances are involved, such as in the case of a long-range cruise missile, a relay platform or satellite may be necessary to pass along the data. In any event the link may be as simple as a way for the weapon to send a final status report in a data burst back to friendly forces; the report from this one-way link might simply indicate that "I'm at this position, doing well, and locked onto the target." It might also include an image of the target as the weapon closes with it.

A more complex, two-way data link arrangement would involve human observation of the emerging target scene and active intervention to cause the weapon to steer or lock on to a specific element within the scene. Such an arrangement requires a much more capable link and offers only a very brief amount of time for an operator to make changes.

In the final functional moment of the weapon, the warhead is activated. In most cases this means impacting the target and detonating the warhead either instantaneously or after a preset delay (to achieve some degree of target penetration). Some warheads, however, are activated without target impact. Submunitions, for example, are dispensed in the air above and short of the target to allow them to disperse properly.

The mission steps that occur after warhead function are discussed in the section "Damage Assessment" later in this chapter. A simplified flow diagram of the basic surface launch steps is found in figure 10.1.

Air Launch Missions

The sequence of steps associated with the air launch of strike weapons bears a strong resemblance to the surface launch sequence, but with some important differences reflecting both the platforms involved and the diversity of weapons

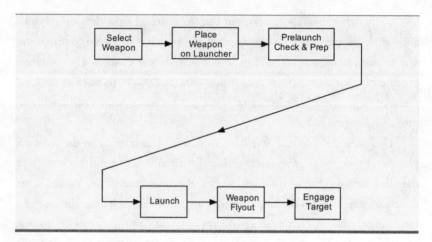

Figure 10.1 Surface Launch Flow Diagram

employed. As before, the sequence begins with the identification of the specific weapons to be used. These must be located in the weapon storage area, which may be a dispersed bunker or open storage at a land airfield, or in an ordnance magazine on board an aircraft carrier. The selected weapons are then extracted from their storage containers and given an initial check to verify that they ready for use. If the weapons are stored in unassembled, component form, they must be assembled before being taken to the strike aircraft.

The weapons are then transported from the storage area to the strike aircraft, where they are attached ("uploaded") to the aircraft. Any final assembly procedures, such as attachment of wings or fins, are accomplished. The attachment hardware is thoroughly checked, a functional test may be performed, and mission data may be inserted prior to takeoff. Any weapons failing to pass final checks are removed ("downloaded") from the aircraft and replaced.

At the appointed time, the strike aircraft takes off. In the case of aircraft carrier basing, this involves a catapult takeoff, referred to as a "cat shot" or "cat." Once airborne, the aircraft navigates the ingress route to the launch point. Aerial refueling may occur during this phase if extended ranges or long airborne loiter times are required for the mission.

As the aircraft approaches the launch point, the strike weapons are activated and final checks are made; final mission data may also be inserted. This initialization process readies the weapons for launch and provides a final check on their status. If a weapon failure is detected at this point, it will usually result in a mission abort.

Depending on the weapon type, the aircraft may have to maneuver considerably to arrive at the necessary launch point. Some weapons use lock on before launch (LOBL) seekers that require the pilot to point the weapon at the target and take some kind of action in the cockpit to lock the seeker onto the target. Other seeker-equipped weapon types use a lock on after launch (LOAL) mode that eliminates this particular step. Still others have no seeker at all and navigate to a set of coordinates that were loaded with the mission data, or, in the case of unguided bombs, they simply fall in a ballistic trajectory.

At weapon release or launch, most strike weapons are pushed away from the aircraft by the attachment mechanism. If the weapon uses propulsion, it is generally started or ignited after a brief delay to ensure that the weapon is clear of the aircraft. The exception to this generality is a forward-fired weapon that is mounted on a launch rail and is accelerated away from the aircraft by its self-contained rocket motor.

The diversity of air-launched strike weapons means that after launch there are a variety of flight path possibilities, depending on the weapon type (fig. 10.2). Unguided weapons simply fall in a ballistic arc. Simple guidance modes incorporated into basic bombs (e.g., LGBs, GPS-guided bombs) result in modified ballistic trajectories. If lifting surfaces are added, they produce more of a gliding flight path. Purpose-built glide weapons tend to flatten the trajectory even more and provide the ability to change course as well as extend range. Powered weapons add range and, in some cases, speed.

If the time of flight from launch to target encounter is substantial, the interim period is called the "midcourse" phase of flight. During this time the weapon is navigating using its own onboard system, often referred to as an autopilot. When the weapon approaches the target, it is said to enter the "terminal" phase. If it uses a seeker, that device generally provides the final steering commands. The target encounter and payload activation for air-launched strike weapons is similar to that for ship-launched weapons.

Data links have been used for quite some time with air-launched weapons. In addition, lasers have been used from airborne platforms to place a "spot" on the desired target element and allow the weapon seeker to home in on that reflected laser energy. Lasers have also been used from nearby ground

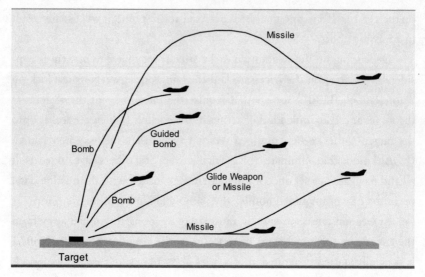

Figure 10.2 Common Air Launch Trajectories

units to designate a target for air-launched weapons, especially during close air support (CAS) missions.

Meanwhile, the launch aircraft depart the area and return to base. Their egress route often differs from their inbound route to avoid retracing their steps through defended airspace. Aerial refueling may once again be used on their way back to base. For those returning to an aircraft carrier, the aviators face an arrested landing, or "trap," when they come back on board.

A summary-level air launch flow diagram is illustrated in figure 10.3.

Damage Assessment

Even with the launch platforms safely out of harm's way and back at base, the mission is not complete until target damage has been evaluated. The usual term for this effort is "bomb damage assessment," or BDA. Since the tasking for the mission usually includes a required level of target damage or destruction, it is important to determine if that has indeed been accomplished.

Several basic issues are usually addressed in this process:

- Was the assigned target hit?
- Was the wrong object hit?
- What level of damage was inflicted?

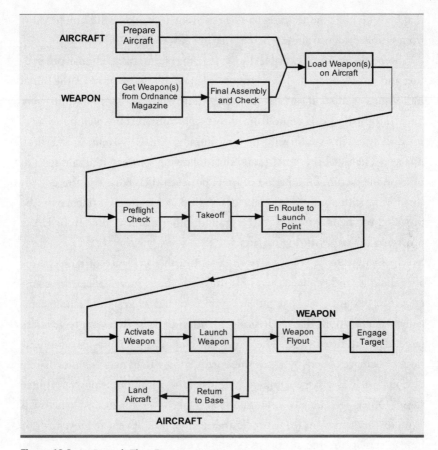

Figure 10.3 Air Launch Flow Diagram

Overall mission performance evaluation also includes some related issues:

- Were any launch platforms damaged or lost?
- Are any crew members injured, dead, or missing?
- Were any enemy air defenses attacked, and, if so, what level of damage was inflicted?
- How much ordnance was expended?

Determining or estimating target damage relies on information from multiple sources. In some cases it is fairly straightforward. If the target was a bridge, and that bridge is now absent, the mission obviously went well. On the other hand, if the target was a small, specific area of a large, robust building, and all that is visible after the attack is a hole in the roof, it would be difficult to know from visible evidence alone whether the attack was successful or not.

Some clues may be available during the time of the attack. Human observation and recorded sensor images from the launch platform are often the initial indications of mission success when they show where the weapons impacted. Such information can come from direct sight, observation of infrared sensor displays such as a forward-looking infrared (FLIR) system, or data link imagery. The usual fire, smoke, and dust following warhead function tend to obscure the picture for a period of time, but even then radar or infrared sensors can provide some additional clues. Additional real-time evidence can be provided by friendly troops on the ground, other manned aircraft, or UAVs observing the target during the attack.

In many cases it is necessary to obtain further evidence of target status after the dust and smoke clears. This usually involves surveillance and reconnaissance resources such as satellites, aircraft, and UAVs conducting data-gathering operations that are basically identical to those necessary to generate target data prior to a strike. The post-strike emphasis, of course, is on indications of damage or activity at the site. Again, some things may be immediately obvious, such as a large smoking hole where a communications relay once stood. Other signs may be more subtle and take more time to evaluate, such as no further emissions from a radio transmitter, lack of vehicle movement or human activity around a supply site, and so on.

The near-term goal of the damage assessment process is to support a decision on whether to consider the target "killed" or plan a second attack. In the longer term a significant number of "dead" targets are kept under surveillance to make sure that any attempts to revive, rebuild, or regenerate them are detected at an early stage. This is especially true of air defense sites or major offensive sites that have been attacked. A regenerated SAM site is almost always considered a priority target, at which point mission planners begin the cycle (planning, execution, and assessment) all over again.

Campaign Strategy

A single strike is but one part of a larger, longer-term campaign during a conflict. The basic approach or strategy used in a campaign is to match the capability of the selected strike weapon with the need that exists at the particular point in a campaign. The warfighters attempt to employ the least complex weapon to satisfy mission objectives.

During the earliest part of a campaign, enemy defenses tend to be at their highest level of capability, and there are generally a number of targets that are considered to be very high value (often because they play a pivotal role in the enemy's warfighting structure). These heavily defended important targets warrant the expenditure of highly capable, long-standoff, and generally very expensive strike weapons, often called "silver bullets." Such weapons are typically in limited supply, so their use is carefully controlled.

After the most critical targets have been attacked and the enemy defenses degraded somewhat, both the relative value of the remaining targets and the capability of the remaining air defenses have declined somewhat. On the other hand, the number of such items is typically many times greater than the number of those earlier, very-high-value, highly defended targets. During this phase of a campaign, large numbers of intermediate-standoff, medium-cost strike weapons tend to be used.

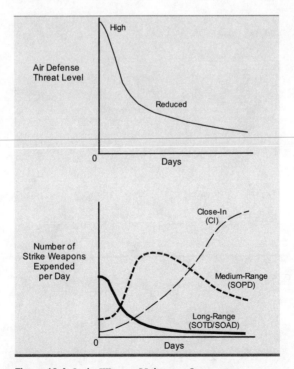

Figure 10.4 Strike Weapon Utilization Strategy

Once the defenses have been reduced to short-range SAMs and AAA, it then becomes practical to use very large numbers of close-in, low-cost strike weapons.

It may be tempting to think of these changes in conditions as distinct phases of a campaign, but, in point of fact, they often exist simultaneously, differing in degree and location. Some areas of a theater of operations are never really heavily defended and thus are open to attack using close-in weapons from the very onset. And heavily defended areas do not just suddenly change character to much-lighter defenses. As a result, it is more realistic to think of overall theater expenditure of strike weapons in terms of a continually changing mix, reflecting the continually changing threat level.

One illustration of this type of expenditure strategy is found in figure 10.4, which depicts the hypothetical use of strike weapons from the three basic standoff classes over a period of time. Long-range standoff weapons dominate the earliest phase, while close-in weapons dominate the later phases, but all three classes are used in varying proportions throughout the conflict.

PART 2
Strike Weapons Development

ELEVEN

Introduction to Part 2

An effective armed force is a combination of competent people and appropriate equipment. If either element is weak, overall capability suffers. Over the years there have been numerous fictional and documentary accounts of the process that changes an everyday citizen into a trained soldier, sailor, or airman. Those accounts have introduced the public to the transformation that occurs to people as they prepare for a far-different lifestyle than they experienced in civilian life. As a result of these various portrayals, civilians generally recognize that there is a difference between the civilian world and the military world, both in wartime and in peace.

However, the same cannot be said for public understanding of military equipment. A limited number of items appear prominently in the media, especially if they represent some major advance in capability, but most items of equipment tend to be far less newsworthy. On the other hand, procurement of military equipment has been the subject of considerable media discussion and much criticism, and has also been the butt of a good many jokes. Accusations of mismanagement, outlandish overpricing, delays, and performance problems are regularly heard. Commentators often paint the subject in negative overtones, leaving the public in a skeptical frame of mind.

Without question, there have been some outrageous examples of each of those faults in military procurements. Both the uniformed members of the armed services who use the equipment and the American taxpayers who foot the bill deserve better. The Department of Defense continues to struggle with those less than satisfactory programs, learning from past mistakes and reforming the acquisition process, but as long as human beings manage the process and make decisions, there are likely to be periodic mistakes or misdeeds.

This portion of the book is neither an exposé, a dreary apology for past problems, nor a starry-eyed, all-is-well account of the subject. Rather, it is an effort to provide readers a factual glimpse into the vaguely reported and often mysterious world of military hardware acquisition, specifically the development and production of strike weapons. Its aim is to provide citizens a better understanding of a very important and unfortunately complex process, again without having to rely on excessive amounts of jargon or technical terms.

What is described here is a summary of development processes that have largely been followed in the United States for the past half century. During that time many of the details and much of the terminology has changed in an evolutionary fashion. This suggests that there are likely to be future changes as well, some perhaps of a more revolutionary nature. With that in mind, the discussion of strike weapons development tends to focus on those fundamental, logical steps that are likely to persist even as regulations, buzzwords, and program structures change in the future.

Hardware Life Cycle

Just about any durable product that we might care to name goes through a series of steps during its life. It begins with a mere notion of something that might fill a need, undergoes some kind of experimentation and development, and then is produced and distributed to users. Often there is an extended period of product support, intended to maintain the item in serviceable condition. And, ultimately, the item is retired from service when it is worn out or replaced by some more modern equivalent.

Military procurement specialists often refer to that sequence of steps as the "womb to tomb" life cycle. Some materials and supplies purchased by the armed services come directly from commercial sources, but these "commercial off the shelf" (COTS) items are not the subject of the discussion here. Instead, we will focus on equipment, specifically strike weapons, that have no commercial application or equivalent and that are the result of the military procurement system.

Like most things military, the formal acquisition or procurement system uses an equally formalized set of sequential steps to define a life cycle. Over time the words and the steps tend to change somewhat, but the underlying themes of the equipment life cycle remain relatively stable. The principal stages are depicted in figure 11.1 and can be described as:

- Requirement: definition of the need
- Assessment: evaluation of alternative ways to satisfy the need
- Development: design, fabrication of experimental and prototype models of the item, test, analysis, rework, and documentation
- Qualification: verification through testing that the item will function properly in the intended environment
- Production: fabrication, acceptance testing, and delivery of quantities of the item
- Deployment and support: distribution of the item to operational units; field support as appropriate; repair and maintenance as needed
- Upgrades: modification of hardware and software to expand capabilities (upgrades are not normally found in the initial program schedule)
- Retirement: removal of the item from operational units at the end of service life; demilitarization as needed

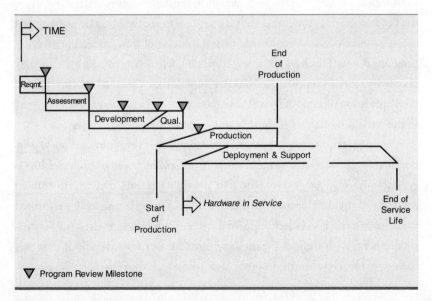

Figure 11.1 Notional Program Summary Schedule

Purists will be quick to point out that this simplified summary does not necessarily match current terminology and that some of the subparts of the process are not explicitly identified. This is true. And it will also be true a few years hence when today's terminology and details change yet again.

Another truth is that there are exceptions to this formalized structure. These generally occur when (1) there is an extreme need for an item, requiring a greatly expedited process; (2) only small quantities of the item are needed; or (3) a novel technology or concept presents an unconventional opportunity during the early stages of the process.

Hardware is considered to be "operational" only during the deployment-and-support stage. Prior to that point it is not yet available to the armed services, even though it may be functioning well in tests. News releases about some new military item being developed can therefore be a little misleading to the public because it may take considerable time before the item is actually deployed.

Program Participants

Strike weapon programs normally involve a variety of organizations during their life cycle. The people who are skilled at defining requirements or evaluating alternatives are generally different from those providing leadership during the production or support stages. It is common to see the program staffing and organizational connections change over the course of time, reflecting evolving program needs. This may seem less surprising when you consider that it may take five to twelve years to go from the initiation of a requirement to the end of qualification of a complex strike weapon, and perhaps twenty to thirty years in the remaining stages of a representative strike weapon life cycle.

Weapon acquisition programs commonly involve a joint military, government, and contractor team. That is to say, military program leadership is typical, with civilian government staff providing technical support. Varying amounts of the development and production efforts are normally performed under contract with civilian industrial firms. Each program tends to be a little different, with different team arrangements and responsibilities, but this three-way mix is the norm. In those cases where hardware items are developed, produced, or supplied by the government as part of the overall program, such hardware is referred to as government-furnished equipment (GFE), government-furnished matériel (GFM), or government-furnished property (GFP). Any data or written material provided to the contractor by the government is called government-furnished information (GFI).

The ultimate customers for all of this effort are the uniformed military personnel who will actually use the end product. Weapon acquisition programs serve the users' needs best when users themselves are appropriately represented in each step of the process.

TWELVE

Security

One of the most basic differences between military equipment acquisition efforts and those of the commercial or private sector is found in the matter of security. As private citizens we all have concerns about the security of our personal and financial information, which causes us to take some basic security measures to protect that information. We keep our records out of the view of strangers, we don't divulge private information to casual inquiries, and we shred material being disposed of that might contain personal data. We also expect the private and governmental institutions that we deal with to protect our sensitive information and our privacy.

Broadening our view a bit, the commercial world certainly imposes specific security constraints on proprietary information, trade secrets, and the introduction of new or novel products. After all, they don't want to give competitors an edge in the business market. Confidentiality of business plans, relationships, finances, and product details is considered normal practice.

Not surprisingly, organizations dealing with military operations or acquisitions have similar concerns about sensitive information and hardware. Over time, a formalized structure has been developed to deal with such security matters, complete with laws, policy documents, implementing instructions, working guidelines, dedicated staffing, and congressional oversight. The details of this formal process do change periodically, but the fundamental considerations and practices remain relatively stable, and these are the subjects of this particular chapter.

Classification Rationale

The only real justification for military security measures is to keep sensitive information out of the hands of potential adversaries. It's basically a matter of keeping the other party ignorant of one's true capabilities, or at least delaying such knowledge for as long as practical. This provides an advantage in time of conflict, when an opponent may not be fully prepared to deal with the capability that you possess.

Starting from this basic objective, various aspects of military and related activities receive differing levels of security protection, reflecting their relative sensitivity. Subjects or objects are "classified" to varying degrees. Those classification assignments are based on written guidelines prepared by senior officials for each project or activity. Appropriate markings are applied to documents or equipment to reflect the assigned classification, and only those personnel who have been authorized, or "cleared," to have access to that classification level are allowed to view, handle, or possess the item.

There are a large number of legitimate reasons to classify an object or subject. In the case of military hardware, there may be design features that need to be protected to prevent an adversary from developing countermeasures against that hardware. In operational matters, it would be foolhardy to openly discuss planned movement of forces or tactics. The specifics of technical research and development efforts are generally restricted to limit an opponent's knowledge of new military technology. Intelligence information is almost always classified to avoid alerting adversaries to the depth of our understanding of their activities and our sources or methods of obtaining such information.

Classification decisions are not taken lightly, nor are they to be used as a screen simply to avoid annoying inquiries from the American public. In point of fact, it is more cumbersome to function in a classified environment than in an unclassified one. Nevertheless, it remains important to protect sensitive material through a rational security system.

Levels of Classification

The federal government has adopted a security system that makes use of several basic classification levels. These levels relate to the perceived degree of damage that unauthorized disclosure would cause to our national defense or foreign relations. Starting with the most sensitive material and working toward less sensitive matters, the three primary levels of classification are defined in government terminology as:

1. *Top Secret.* National-security information or material that requires the highest degree of protection, and the unauthorized disclosure of which could reasonably be expected to cause exceptionally grave damage to our national security. Examples of "exceptionally grave damage" include armed hostilities against the United States or its allies; disruption of foreign relations vitally affecting national security; the compromise of vital national defense plans or complex cryptological and communications intelligence systems; the revelation of sensitive intelligence operations; and the disclosure of scientific or technological developments vital to national security.

2. *Secret.* National-security information or material that requires a substantial degree of protection and the unauthorized disclosure of which could reasonably be expected to cause serious damage to our national security. Examples of "serious damage" include disruption of foreign relations significantly affecting national security; significant impairment of a program or policy directly related to national security; revelation of significant military plans or intelligence operations; and compromise of significant scientific or technological developments relating to national security.

3. *Confidential.* National-security information or material that requires protection and the unauthorized disclosure of which could reasonably be expected to cause damage to national security.

These three basic classification levels are supplemented by several other classification terms that exist below the Confidential level. The most obvious lower level is "Unclassified," meaning that the material is judged to be not sensitive. However, there are other terms used within the government to indicate that while the information may not be sensitive from a national-security point of view, it may be sensitive for other reasons, and therefore not automatically releasable to the general public. Such terms include "Privacy Act information" (related to personnel files) and "for official use only" (often related to internal policy matters).

To avoid misunderstandings classified writings are always marked to indicate the level of classification of the information being discussed. Individual paragraphs are marked at their beginning to indicate the highest classification of information in that paragraph, using a (U) to indicate Unclassified, (C) to indicate Confidential, (S) to indicate Secret, and (TS) to indicate Top Secret. Each page is also marked at the top and the bottom with the complete word of the highest classification level of anything appearing on the page. Likewise, the document cover is marked to indicate the highest level of classification of anything found within the document.

In terms of the volume of information marked, the vast majority of military and related documentation is at the Unclassified or Confidential level. A much-smaller volume of material is found at the Secret level, and an even smaller amount is at the Top Secret level. Over time, classified materials are reviewed for possible downgrading to lower levels, and when appropriate, such documents are re-marked accordingly.

Clearances

The people who actually work with classified materials must undergo background investigations to verify that they are sufficiently trustworthy to have access to such sensitive information or equipment. As you might suspect, the extent of the investigation varies with the level of classification access that the individual is seeking. A clearance for the Confidential level is the least extensive, while a clearance for Top Secret is lengthy and goes into great depths.

It is fairly common for individuals to obtain an initial clearance at a modest level and later undergo an additional investigation for an upgraded clearance level. Clearances are reviewed periodically and follow-up investigations are routine, particularly at the higher clearance levels. Behavioral or lifestyle changes may also trigger follow-up investigations. Random drug testing and even polygraph (lie detector) examinations may be used by federal agencies as part of their continuing security-screening process.

Any hopes of personal privacy must be abandoned for those holding Top Secret clearances. All aspects of their lives are open to scrutiny by security examiners, including what is normally considered their "private lives": that is, their history, law enforcement records, legal actions, personal finances, family, interests, personal contacts, travel, and so on. The information collected by security investigators and examiners is protected from casual access by the general public, but it remains part of the individual's personnel records in a separate file.

Need to Know

Classification levels and corresponding personnel clearances are just one part of the national-security system. Another basic tool that is used to protect sensitive material involves a concept called "need to know." Individuals who are involved in some way with the particular subject are granted access, since they have a legitimate need to know about the matter. Other individuals who may have the proper clearance level but who have no particular role in the matter may be denied access under the need-to-know process. The more sensitive the information, the more certain this practice will be followed. The underlying reason for the practice is to limit the number of people who have access to sensitive material and thereby reduce the possibility of inadvertent disclosure.

Special Access

An even more rigid form of the need-to-know philosophy is the practice of compartmentalization. This is a formal process involving use of separate access identifiers in addition to classification-level designations. Only those individuals specifically cleared for the particular subject or compartment are allowed access.

Some of the access identifiers that have been used in this process include special access required (SAR), special need to know (SNTK), sensitive compartmented information (SCI; for certain intelligence information), Restricted or Formerly Restricted Data (referring to early nuclear weapons information), NATO (for certain information shared jointly with other NATO participants), and Critical Nuclear Weapons Design Information (CNWDI). Additional identifiers or code words may be used to further define the specific compartment. A well-known example would be the term "Have Blue," which was used as the code for the early development work that led to the F-117 stealth fighter.

In practice, documents are marked with both the classification level and the special-access identifier. For example, an early stealth fighter technical document might have carried the marking TOP SECRET/HAVE BLUE.

A few topics or projects are so highly classified that even their code words are classified. No public acknowledgement is made by the government that such an effort exists. These are often referred to in the public press as "black" projects, meaning that all information about them has been blacked out. In a number of cases, later revelations made by the government have noted or described such efforts after the associated level of sensitivity had diminished to an acceptable level. The Tacit Blue project involving an experimental aircraft built by Northrop and tested by the U.S. Air Force would be an example of a black project that was conducted under the utmost secrecy but later openly discussed.

Other Caveats

Government or military documents may also carry other restrictive markings, limiting access in one way or another. Materials related to contractor proposal evaluations by the government may be restricted to government personnel only. Intelligence materials may be restricted to U.S. personnel only and carry a NOFORN/WNINTEL (no foreign/warning notice—intelligence sources and methods) marking. The NOFORN label may also be applied to highly sensitive technologies, especially during their early development. When preliminary information is first made available for limited review there may be justification to prevent its further circulation, and it may carry an ORCON (originator control) marking to ensure that nobody but the originator makes additional distribution. And, of course, we're all familiar with the EYES ONLY marking that figures prominently in spy novels, indicating extremely limited distribution to only a few named individuals.

Facility Security

Secure storage and handling of classified materials requires more than proper marking and clearances. It also requires appropriate facilities and procedures, aimed at safeguarding the sensitive materials both while they are in use and when they are inactive. The underlying objective of this aspect of the matter is the prevention of unauthorized access to sensitive items.

It should be obvious that just about any organization has material that is nobody else's business or that might tempt a thief. It is therefore common to see signs posted around business offices or establishments that read Employees Only, Authorized Personnel Only, or Private. Such facilities also have locks on at least some of the file cabinets, and they lock their doors at the end of the business day. Many businesses have even gone to the trouble of installing burglar alarms or have employed surveillance cameras or night watchmen as loss-prevention measures.

Security measures at facilities housing classified materials are simply an expansion of commercial practices. Once again, protection requirements escalate in proportion to the level of sensitivity of the materials held on-site. If nothing higher than Confidential material is to be held at the facility, the security requirements are relatively modest. However, if Top Secret material is involved, security requirements are quite extensive.

In all cases facility security is based on the establishment of barriers to intrusion or unauthorized access. The classified material itself (document or hardware) gets secured behind a lock when not in use. Office safes with combination locks are commonly used to store classified documents; locked containers or spaces are used for classified hardware. When the material is out of storage, it is used in a secured area to prevent unauthorized personnel from observing anything classified. This generally means that workspaces are behind locked doors or that access to the space is controlled by guards.

As classification levels increase, so do the necessary security measures implemented at work facilities or test sites. In many cases common electronic devices such as cell phones, PDAs, flash drives, and other items that can either store or transmit information are simply banned from the workplace. Both electronic and paper communications leaving the facility have to be handled carefully to avoid inadvertent disclosure of classified information. Laptop computers that leave the site are a special concern, since it is all too easy for classified material to end up on them by mistake. Even the background electronic "noise" of desktop computers within the secure facility may be an issue unless there is insufficient shielding in place to prevent those signals from being received off-site.

Even with just a cursory look at the security practices involved in military matters, it is obvious that a great deal of effort is put into protecting sensitive information and hardware. Such effort is costly and cumbersome, and results in an added burden when compared to commercial practices. However, in the context of national security in a troubled world, it is unfortunately necessary. To do otherwise would be to jeopardize the men and women of our armed forces and ultimately our civilian population and institutions as well.

THIRTEEN

Defining the Need

If product development is all about filling a need, then a clear understanding of the need is essential. When a person decides that he or she needs a motor vehicle for transportation, a thought process is begun to help determine what kind of motor vehicle would be suitable. Popular advertising may promote all sorts of automotive alternatives, but a thoughtful individual will try to establish some guidelines before visiting dealer showrooms: How will I use the vehicle? How large a vehicle do I need? What are my budgetary constraints? And so on.

Similar questions and more are asked during the course of defining the need for military equipment. The "requirements definition process," as it is generally called, reflects the urgency, magnitude, and complexity of the particular item or capability in question. Some items are straightforward, well understood, and mature; these proceed through the requirements process very quickly. Others may involve capabilities that require new technology or very complex interactions; these tend to take a bit longer to adequately define.

User Requests

One of the several ways that a requirement is initiated is via a request from the operational forces. A concern or complaint expressed by deployed units generally gets prompt attention, especially if there is a perceived safety or security risk. The forward units may have encountered a deficiency in existing hardware, or they may be facing something unexpected from the adversary.

Their call for help may be addressed in more than one way. Sometimes it is possible to resolve the issue with a quick fix or a short-term solution. Emergency repairs of faulty parts would be one example. Adaptation of existing hardware would be another. An example of the latter situation would be the Navy's quick-response Shrike On Board (SOB) project in the 1970s.

When American destroyers approached the coast of North Vietnam to conduct shore bombardment operations, they were taken under fire by well-aimed, radar-directed shore batteries. The Navy quickly assembled a small team to adapt the air-launched AGM-45 Shrike antiradiation missile to surface launch, something that had never been done nor anticipated. Within days the team had created a simple launcher, mounted it onto an existing piece of gear on board ship, and verified that Shrike could indeed be launched from a surface platform, even in high crosswinds. A limited number of launchers and related support equipment items were fabricated and installed on selected vessels in the Pacific. Once SOB was deployed and in operation, the threat from radar-directed shore batteries was all but eliminated. Total time from the request for help to the first operational launch off Vietnam was 104 days. As a short-term solution to a short-term problem, SOB was retired from service after a relatively brief operational period.

Other user requests may require much more in the way of development of a new capability. The expressed needs may imply significant improvements in current capabilities, well beyond what a simple modification might produce. These requests initiate a more rigorous and structured requirements definition process, involving a wider circle of participants.

Threat Projections

A second major source of requirements initiatives comes from the combined operational and intelligence community. This involves estimating the capabilities of potential adversaries, comparing those with our own capabilities and determining what deficiencies may exist. This is done for both current situations and for projected future conditions from five to perhaps twenty or thirty years in advance.

Obviously, the farther into the future you project, the less certainty exists. Threat projections made in the mid-1980s for the Soviet Union did not anticipate its collapse and the subsequent restructuring that followed. Similarly, the focus on a relatively stable, high-technology adversary during the Cold War era did not fully address what might be needed to cope with poorly defined insurgents and widespread terrorism.

Despite the unstable nature of global politics, a close coupling of technical intelligence findings with operational planners remains an important source of emerging equipment requirements. Projections are based on the best available information at the time. Waiting until a conflict occurs to start the process would cause the response to arrive too late.

Technology Opportunities

A third major influence in the requirements process is the technology community. In one role, technology establishes some current boundaries or limits on what can reasonably be accomplished; that is, the current state of the art allows you to do this much but no more. A good understanding of the applicable technical fields should also help identify the level of risk involved in attempting to expand the current state of the art.

In another role, and perhaps the one that gets more media attention, technology advancements or breakthroughs may offer unique opportunities to greatly alter military capabilities. The guidance technology that first allowed truly precision guided weapons is one example. Miniature jet engines for cruise missiles is another, as is stealth technology. When an important technological advancement comes along, its promise may be sufficient to initiate a new military requirement. At the same time military planners must be careful not to push some clever new gadget without having a genuine need for it; this situation is often called "a solution looking for a problem."

Since strike weapons are expendable items, intended for onetime use, technology that supports simplified, affordable hardware is always of interest. This kind of technology seldom makes the headlines, but it plays a very important role in providing our armed services the quantity and quality of hardware that they need.

The Formal Requirement

Regardless of the originating source of the expression of need, some kind of formal requirement statement is ultimately written down. The form and format of such statements varies with the military services and with the era. During

World War II, for example, one particular requirement for a new general-purpose bomb for Navy dive bombers was defined in less than one page of paper (the rest of the page described the required proof testing and the initial production quantities). Today, in spite of DoD instructions to be brutally concise, a basic requirements document for a new strike weapon is usually several pages in length.

The requirement document itself is prepared by a designated office or staff within the particular military service. Inputs may be obtained from various sources, solicited or otherwise, but the ultimate responsibility for requirement preparation is retained by the military service.

The intent of the formal requirement is to identify those capabilities, features, attributes, and program specifics that are absolutely essential to meet the needs of the user. A proper requirement statement lists only those things that are truly required, and avoids "nice-to-have" options. The language should be clear and direct, without vague or ambiguous wording. When complete and signed off, the requirement document can be thought of as a contract between the operational forces who need the capability and the acquisition community that will develop, qualify, and produce the item.

A number of topics need to be addressed in abbreviated form in a strike weapon requirement document. These include the following:

- *Basic mission.* This is a brief statement of the primary mission (e.g., short-range battlefield interdiction).
- *Launch platforms.* These would be identified by general category (e.g., air launch, surface launch) and specific required platforms (e.g., B-2, F/A-18, etc.).
- *Targets.* Again, these would be objects identified first in general terms (e.g., armored vehicles, parked aircraft) and then by specific examples as appropriate (e.g., T-90 main battle tank, Su-27 aircraft). When a minimum required level of target destruction is required, it is often identified in this section (e.g., 0.7 probability of mobility kill with single shot [Pk is always expressed as a decimal fraction; 1.0 is a theoretical upper limit that is never achieved]).
- *Threat category.* This identifies the level of target defenses that the weapon must be able to cope with while providing a reasonable probability of launch platform survivability. This may be expressed in general terms or

by inclusion of specific air defense systems. Realism must be applied here to avoid inappropriate combinations of mission/platform/target/threat level. For example, calling for a short-range weapon for use in a very-high-threat zone is likely to place the launch aircraft at high risk, while asking for a long-range weapon to be used in a low-threat zone seems a waste of an expensive asset.

- *Operational envelope.* Key aspects of the required launch envelope are listed (e.g., maximum and minimum launch range, maximum and minimum launch altitude for air-launched weapons, off-axis launch requirements, minimum and maximum platform velocity at launch, etc.). Minimum essential operating conditions are also listed (e.g., usable day or night, in fog, clouds or light precipitation). Prelaunch conditions on the platform are often listed (e.g., carriage to the full subsonic operating envelope of the specified aircraft).

- *Physical envelope.* Maximum weight and dimensions of the weapon are usually related to handling and launch platform criteria. However, it is not uncommon to see overall limits identified in the requirements document to ensure that certain thresholds are not exceeded. These might include maximum allowable gross weight and limits on overall length or weapon cross-sectional size.

- *Availability.* The requirements document often identifies a required initial operational capability (IOC) date to establish the time frame that the weapon is needed by the operational forces. This helps focus the development effort on those alternatives or technologies that will be sufficiently mature to support the program.

A certain amount of analysis is generally needed to arrive at a realistic and balanced requirements statement. Experienced staff at the appropriate service command can often accomplish the initial requirements balancing in a relatively short time for well-understood items. But the process can be extended considerably when new technology is involved or when multiple launch platform categories are specified. And once the "technical" balancing has been completed, there still remains an often lengthy review chain before the requirement has been coordinated, signed off on, and approved. Not all requirements make it through the process, and some that ultimately are approved may take several years to reach that point.

FOURTEEN

Evaluating Alternatives

Once the requirement has been approved, the next step in the acquisition process is to search for alternative ways to meet it. Not all weapon requirements lead to completely new hardware. Sometimes it is possible to satisfy the need with modifications or growth variants of existing systems. Or perhaps some other service or friendly nation has something suitable that could be procured. This next step is intended to explore multiple alternatives before reaching a conclusion on how best to respond to the new requirement.

Analyzing the Question

The people who are tasked to evaluate alternatives are generally not the same ones who generated the requirement. It is therefore important that some initial time be devoted to studying the requirement so that it is completely understood. Liaison with the requirements staff is generally accomplished during this process, to ensure that there is no ambiguity of terms or misinterpretation of specific needs. While this step may be relatively brief, it is absolutely vital to a successful outcome. Unless the evaluation team truly understands the question, there is little reason to believe that it will arrive at a proper answer.

Concept Formulation

This stage of the process usually involves the services of experienced system engineers within the military service organization who can rapidly formulate basic technical concepts that would address the needs found in the requirement. Their notions may be supplemented with ideas from private industry, often solicited through a mechanism called a request for information (RFI). The very first working sessions during this phase are sometimes characterized as brainstorming sessions, where there is a free exchange of ideas, some of them exotic or seemingly far-fetched. The goal at this point is to identify a large number of potential solutions, be they brand-new designs, modifications of existing systems, or procurement of items from sister services or friendly foreign nations.

Initial Screening

These initial concepts are then compared to basic operational and physical parameters found in the requirement document. For example, if the requirement document indicates that the weapon must not weigh more than a certain amount, any concept that is substantially heavier will be screened out. Likewise, a need for all-weather capability would screen out concepts that would only work during day, clear-weather conditions. This first screening generally eliminates a significant percentage of the initial concepts and allows the evaluation team to focus attention on a limited number of alternatives that actually have the best chance of satisfying the requirement.

Evaluating Viable Options

Detailed evaluation of screened alternatives takes on a more formal tone. Such efforts generally follow specific guidelines set down by the particular service and the Department of Defense. The terminology changes from time to time, but such efforts have been identified as cost and operational effectiveness analysis (COEA) or analysis of alternatives (AoA). Terminology aside, the fundamental purpose remains the same: a side-by-side evaluation of the estimated operational capabilities and life cycle costs of each of the screened alternatives. To use Pentagon vernacular, the evaluation is intended to estimate the amount of "bang for the buck" for each program option.

A variety of analysis tools may be employed in the detailed evaluation. The concepts first have to be sufficiently defined to allow their performance and operational characteristics to be estimated. Conventional preliminary-design tools and methods are generally used for new concepts, while actual data should be available for existing systems.

Estimates also need to be made of what are called programmatic features for each alternative. These elements include estimated development and qualification programs, production runs, and the like. Alternatives that are basically the procurement of existing hardware will exhibit far-shorter schedules to first operational capability than those requiring full development efforts.

Cost estimators combine the technical and programmatic material to forecast the financial outlay for each alternative. Costs are estimated for each major phase of the program anticipated for each of the alternatives. Final life cycle cost estimates are projected by including development, qualification, production, operational, and support costs. When several thousand production units are involved, the common focus on initial development costs can be greatly overshadowed by production and support costs. The cost of launch platform modifications, launchers, support equipment, and other needs must also be included in life cycle cost estimates.

Operational-effectiveness analysis is handled by a separate team of specialists who make use of various computerized tools to evaluate the alternatives' capabilities in combat situations. Basic scenarios are created for the anticipated time frame of interest, reflecting the theme of the weapon requirement and best estimates of the kind of threat that would exist at that time. The several alternative weapon concepts are then individually employed in that threat environment to estimate both losses of friendly launch platforms (and strike weapons) to the defenses and the effects of the strike weapons against the designated targets. Experienced military operators may be part of the strike-planning process to ensure that rational tactical employment decisions are made. Variations in the amount of strike support may also be analyzed as part of the trade-off process. The underlying goal of this effort is to produce an "apples-to-apples," direct comparison of the strike effectiveness of the several alternatives.

The above brief overviews of the several major steps in the evaluation process mask what is often a large and complex effort for major weapons programs. It is essential that the process be carefully and honestly managed to avoid biases or influence exerted by proponents of a particular alternative. It is equally important that risks and uncertainties be identified, especially with unproven concepts or new technology, so that the military service or Department of Defense decision makers can be fully aware of such issues.

Selecting a Course of Action

The evaluation of alternatives leads to a decision made by senior staff. The decision may be to stop and proceed no farther, or it may be to select one of the alternatives. The results of the analysis often bring new understanding to the overall matter, which may lead to revision of program goals or features. In most cases a program that has progressed to this point will continue on to the next stage, but only after a high-level review within the service or the Department of Defense.

Sometimes the assessment leads to a conclusion that additional research is needed before one or more elements of a suitable concept are ready for full-scale development. This may result in a decision to add a predevelopment effort, often called a demonstration phase or a validation phase, to the overall program plan. The purpose of such activity is to improve the state of the art in critical areas and to verify that all elements of the system concept are indeed sufficiently mature to warrant entry into development. This preliminary phase would typically involve some level of testing and would end with a program review, during which the decision makers would again consider overall status before determining whether it was appropriate to proceed into development.

For purposes of this book, it will be assumed that there has been a favorable decision, with or without a predevelopment effort, and that the program is moving forward.

FIFTEEN

Design Criteria

Before actual development work begins on a new strike weapon system, the newly appointed program staff needs to generate some key materials to guide the development. On large programs the amount of institutionalized paperwork can be staggering as program plans, schedules, test plans, work organization charts, contractual documents, and various other bits of administrative minutiae are drafted, coordinated, reviewed, revised, and published. Even small projects are often heavily burdened with significant paperwork requirements, although not generally to the extent of their larger counterparts.

Setting aside the many programmatic documents, there is also a critical technical document that gets prepared: the system specification. This can be thought of as a technical expansion of the very basic material found in the earlier requirement document. It provides added clarity and definition of essential technical and operational parameters for the people who will subsequently design, develop, and qualify the new hardware. When development work is contracted out, as is often the case, the system specification is a key part of the contract terms.

System Specification

The system specification defines the complete weapon system in functional and physical terms. The relationships between the various system elements are described, along with their interfaces. For example, a missile system specification would include discussion of the missile itself, the launcher mechanism, launch control equipment, launch platform(s), support elements, and the like.

Physical and functional criteria are noted for each element or subelement, relevant military specifications (MIL-SPECs) and military standards (MIL-STDs) are cited, and performance levels are identified. In short, the system specification (or "spec") is supposed to be a thorough description of the capabilities of the desired product, written in a way that will enable a competent technical team to independently develop that product.

While the operational-requirement document would typically be no more than a few pages in length, a system spec can easily exceed fifty pages and can sometimes run several times that length. This is especially true of complex weapon systems that may have wide application.

Spec language can seem strange to the uninitiated. The word "shall" takes on special meaning in such writings, as it indicates a contractually binding, nonnegotiable feature or characteristic. Technical parameters often seem to be mixed with legalistic terms. Some aspects are laboriously described in words, while others may be delineated using figures or tables that identify requirements or limits.

Because a system specification is a formal program document, there are institutional guidelines on how to prepare one. Some of the particulars have changed over time, but the basic intent of the document has remained stable, as has the basic organizational structure. The document has six mandatory sections, with all but the final section ("Notes") being contractually binding:

1. SCOPE
2. APPLICABLE DOCUMENTS
3. REQUIREMENTS
4. VERIFICATION
5. PACKAGING
6. NOTES

A certain amount of flexibility is permitted in writing system specifications for the wide variety of military hardware that is procured by the Department of Defense. As a result, the contents of the several mandatory sections can vary from system to system. When it comes to design criteria, most of the attention is focused on section 3, "Requirements," which is further expanded into the subsections noted below.

3. REQUIREMENTS

3.1 System Definition

 3.1.1 General Description

 3.1.2 Mission

 3.1.3 Threat

 3.1.4 System Diagram

 3.1.5 Interface Definition

 3.1.6 Operational and Organizational Concepts

3.2 Characteristics

 3.2.1 Performance

 3.2.2 Physical Characteristics

 3.2.3 Reliability

 3.2.4 Maintainability

 3.2.5 Availability

 3.2.6 System Effectiveness

 3.2.7 Environmental Conditions

3.3 Design and Construction

 3.3.1 Materials, Processes, and Parts

 3.3.2 Workmanship

 3.3.3 Interchangeability

 3.3.4 Safety

 3.3.5 Human Interfaces

3.4 Documentation

 3.4.1 Drawings

 3.4.2 Specifications

 3.4.3 Technical Manuals

 3.4.4 Test Plans

 3.4.5 Configuration Management and Control

3.5 Logistics

 3.5.1 Maintenance

 3.5.2 Supply

 3.5.3 Facilities and Equipment

3.6 Personnel and Training

3.7 Functional Area Characteristics

3.8 Precedence

Brevity is a virtue in spec writing, but the topic still must be sufficiently addressed to permit an unambiguous interpretation of what the customer wants. Some portions of the document can be quite brief, while others tend to become lengthy. In all cases only essential, must-have parameters are included. The system spec is not a place to include all sorts of optional or nice-to-have features.

Essential Characteristics

From an operational point of view, most of the items of primary interest are found in sections 3.1, "System Definition," and 3.2, "Characteristics." These are also the areas that seem to be of most interest to the public, because they deal with the most popular topics: mission, threat, operational concept, performance capabilities, size, weight, effectiveness, and so on.

For the most part the parameters found in the system spec are derived directly from the earlier operational requirement. The spec provides further elaboration as necessary on such things as mission, threat, and operational concept, but it does not alter the fundamental requirements. Similarly, parameters such as maximum size and weight are derived from the need to be compatible with the launch platforms identified in the operational requirement.

Performance and effectiveness criteria often are the subject of considerable analysis before they are added to the system spec. The summary needs expressed in the operational requirement are expanded into a rational and more complete definition of technical parameters in the spec. This portion of the document tends to become more sensitive and classified than many other parts of the system spec because it deals with very important operational capabilities: for example, launch envelope, standoff, speed, maneuverability, survivability, and warhead effects.

Operational-Environment Appendix

One of the larger subsections in a weapon system spec would be the discussion of the operational environment (section 3.2.7 in the above example). Because of the depth of definition often needed for that topic, it is common for the material to be documented separately as an annex or appendix to the basic system spec.

Perhaps the biggest difference between commercial equipment and comparable military equipment is the latter's need to reliably and safely function after exposure to some very harsh environmental conditions. We don't really expect a common, personal laptop computer to continue to work after being dropped several feet onto concrete, blasted in a sandstorm, exposed to soaking saltwater spray, left out overnight in temperatures well below zero, and zapped with high-energy microwave radiation. But someone in combat providing target information to artillery or strike weapon shooters might just need a laptop that will withstand all that and continue to function properly. Their lives might depend on such robust equipment.

Defining the environmental-design criteria for a weapon system is not a matter of simply listing a series of MIL-STDs or MIL-SPECs. Instead, it begins with a careful examination of the anticipated life cycle of the equipment, from the moment it comes off the production line to the moment it is expended or retired from service. It considers each step in the sequence of the life cycle, including storage, handling, transportation, deployment, and operational mission segments. Each of those is further broken down into logical condition stages, where the equipment might be inside a container, exposed but not activated, activated, and so on. Then environmental-condition data are applied to each step to define the type of stimulus likely to be experienced.

An air-launched guided missile, for example, might leave the factory inside a container and be transported by truck or rail to a military depot facility, where it could be kept in a storage bunker for an extended period. It then might be moved to outside storage for a short period, awaiting further transportation. The next stage might involve transportation by ship to some forward operating area, where it would be stored, still in its container, in an open ammo dump. At some point the container could be opened, permitting the missile to be checked and then brought to a staging area. Subsequently the missile could be uploaded onto a strike aircraft and carried on the wing of that aircraft for one or more flights in an unpowered status. Ultimately, on an ensuing flight, the missile is powered up, initialized, and launched, whereupon it makes its way to the assigned target.

While this example sequence is fairly simplistic, it does point out that the hardware is likely to be exposed to a variety of environmental conditions along the way. Variations occur in both length of time and intensity of exposure to stimuli such as temperature, pressure, humidity, solar radiation, precipitation, vibration, acoustics, acceleration, physical shock, and structural loads, among others. Additional military environmental factors include exposure to blast waves, electromagnetic radiation, saltwater spray, sand and dust, fungi, launch platform fuels, and hydraulic fluids.

Even if you ignore the rigors of foul weather and combat operations, equipment items carried by military aircraft face some challenging environmental conditions. Storage conditions may vary from the frozen arctic to blistering deserts. When carried on the aircraft, the item may experience ambient pressures and temperatures associated with altitudes from sea level to extreme heights. The routine flight environment includes air loads, vibration, and noise, along with forces caused by aircraft maneuvering. All of these environmental conditions and stimuli need to be clearly and rationally defined in the operational environment section of the system specification.

As individual pieces of the weapon system are broken out for development, their functional and physical requirements must be carefully extracted from the system spec, along with their associated environmental requirements. Equipment developers may sometimes be tempted to set aside certain military-grade environmental conditions for "minor" interior parts in hopes of using inexpensive commercial equivalents, but these expectations are often dashed when later qualification testing uncovers a weakness or failure due to the use of such parts.

Safety Issues

Strike weapons are intentionally dangerous. They are intended to inflict great harm and destruction to enemy targets. Simultaneously, their developers would prefer that they not pose large risks to the friendly forces that employ them. These seemingly opposing goals lead to some challenging safety design criteria.

The warhead components of the strike weapon receive special attention with regard to safety requirements. Prior to being launched against a target, the complete ordnance package must be able to withstand very harsh treatment without reacting in a deadly manner. For example, it must be capable of remaining safe during and after a forty-foot drop onto steel-covered concrete. Similarly, it must not detonate after being hit with gunfire or after being subjected to an enveloping fire fed by jet engine fuel. These requirements have been in place since the late 1960s and have indeed been met by American strike weapons developed since that time, a rather remarkable achievement.

Stringent safety requirements also apply to the mechanism that arms the warhead fuze. It must be exceptionally reliable and must allow arming to occur only after onboard sensors have determined that the weapon was indeed intentionally launched. This allows weapons to be jettisoned in an unarmed (safe) state should the need arise.

At the same time, the warhead must also function reliably when it encounters the target. In many cases this means designing components to survive extremely high shock loads on impact with a sturdy target to allow for some amount of target penetration before the warhead detonates.

Design Trade-Offs

The field of engineering frequently employs trade-offs and compromises to accommodate conflicting needs in desired attributes. The desire for long stand-off, for example, must be balanced against the associated weight and volume penalty for propulsion components and fuel. A goal of extreme accuracy may only be achievable with a very precise and expensive guidance system. High survivability may require signature control methods that complicate the design.

The fundamental question in design trade-offs is this: At what point has the design achieved an acceptable balance between the various performance goals? There is no firm rule or guideline to use in answering this question. It becomes a matter of judgment, hopefully an informed and rational engineering judgment rather than an arbitrary or capricious decision. This is an area in which experience is very valuable, placing the more senior system engineers in a dominant role.

Strike weapon design trade-offs are further complicated by their "one-shot" operational application. That is, they are intended for use once, and during that use they will be destroyed. They are not designed for repetitive operation, as might be the case with combat aircraft, a ship, a tank, or even an assault rifle. Strike weapons are more like a cartridge for the rifle—an expendable item used once. Granted, strike weapons cost a tremendous amount more than cartridges, but they are still single-use items.

The expendable nature of strike weapons adds an important design trade-off dimension to the development and production process—cost. Weapons must be both effective and affordable. The design process therefore includes cost implications at every level, again using a trade-off philosophy that balances expense against performance. As with other trade-offs, the experience of the design team is important to making good trade-off decisions in this area as well.

SIXTEEN

Development

The size of the program workforce expands dramatically when development gets under way. This is the program phase that the media most closely associates with the term "research and development" (R&D), or its more complete form: "research, development, test, and evaluation" (RDT&E). In actual fact most of the research should be complete before reaching this point, but the popular terminology lingers anyway.

Development almost always requires a major outlay of resources. The decision to enter development is therefore a very critical milestone in a program, and one that is carefully reviewed before permission is granted by high-ranking officials. The decision to do so carries with it a commitment to provide the necessary funding, staffing, and facility resources for an extended period of time. If there is any doubt about the necessity for the item, the wisdom of the proposed acquisition plan, or the availability of resources, this is the time to halt the process.

Once a decision is made to proceed into development, activity accelerates and expands. The internal team is formalized, contracts are prepared and let, and all efforts are focused on turning the proposed strike weapon concept into reality.

Phased Activity

The breadth of activity during development reflects the breadth of the system spec. For a brand-new weapon system, work occurs on a great many individual aspects: the weapon itself and all of its component parts, any launchers or launcher modifications, the various launch platforms, supporting elements, and so on. Both hardware and software must be developed, and all elements must be properly integrated into a reliable, safe, and smoothly working system that is suitable for use in a harsh combat environment by uniformed service personnel. Something that works well when operated by a skilled technician in a pristine laboratory environment may not necessarily be suitable in the field of battle. The realities of the end use of the system components must therefore always be kept in the forefront.

A simple diagram may help illustrate the manner in which multiple elements typically found in a weapon development program all come together. Figure 16.1 begins that buildup on the left with various parts and small components. These are assembled into a number of subsystems, which then make up the final system. This particular diagram has been kept quite simple to illustrate just the basic concept. The actual buildup diagram for a real weapon system would be very large and complex, reflecting the thousands of small parts and components involved.

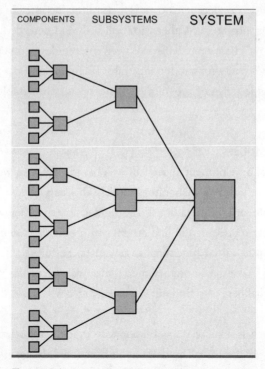

Figure 16.1 Generic Development Progression

The left-to-right orientation of figure 16.1 is basically an engineering representation of the flow path from small pieces to larger subsystems to the final system. If the diagram is rotated counterclockwise to form a pyramid, you have a commonly used management diagram for the program. Invert the pyramid and you have a documentation tree, in which specifications and drawings for subsystems and components expand and flow upward from the system spec.

Because there generally is such a wide range of elements being pursued in parallel, virtually all strike weapon development programs place heavy emphasis on constant collaboration between the various subteams and a step-by-step or phased approach to the combined efforts. The program terminology and phasing criteria change over time, but the fundamental approach is actually quite straightforward. The development program methodically goes from a paper design to a laboratory model to a working prototype to a production-ready item in a series of steps.

Some parts of the system will be relatively mature, meaning that they will require very little development work but still must be integrated with other components or subsystems. Other parts will be brand new, perhaps based on new technology. These new items will need intense development activity to bring them to a ready-for-production state. Balancing and coordinating the efforts for a wide range of program elements takes considerable skill on the part of the program staff.

LABORATORY MODELS

A conventional development approach for a brand-new item will begin with a first attempt at building a laboratory model. This step may be called various names: a breadboard model, a brassboard model, an experimental model, and so on. It simply represents the first attempt to mechanize the concept being developed so that initial functional tests can be performed. Often there is no attempt to make the laboratory model meet any particular size or weight criteria. The goal here is to find out if the basic idea will work. Packaging will come later.

There often is a succession of laboratory models, or at least significant modifications to the initial model, as the functional design is worked out. Once confidence has been established that the component is operating satisfactorily at this stage, the development effort expands to incorporate packaging, integration, and operational-environment issues.

ENGINEERING MODELS

The second basic phase of a development effort is focused on what are usually called engineering models. This is when items are brought together to begin functioning collectively. To do so, the components need to fit within their assigned volumes and must communicate with each other in the agreed-upon manner.

Engineering models of various configurations undergo considerable testing. Some of the testing retains the laboratory atmosphere as components and subsystems go through more ambitious examinations of their performance capabilities. The aerodynamic characteristics of the overall configuration are typically evaluated by both analytical tools and by wind tunnel testing. Ordnance components are evaluated in arena tests, during which the explosions and their results can be examined in detail. If the warhead is required to penetrate the target before detonation, penetration tests will be run using rocket-propelled test articles launched from test tracks.

Flight testing follows an evolutionary pathway as well. Portions of the engineering models, such as guidance subsystems, may be carried aloft by aircraft so that they can be tested in representative flight conditions; this practice is called captive-flight testing, meaning that the parts remain captive on the aircraft and are not actually launched.

Free-flight testing involves the release or launch of the weapon from the launch platform. The first such tests use dummy shapes that are of the proper configuration and weight but contain no active internal components. These are used to verify that a safe release or launch can be accomplished under various launch conditions. The second usual free-flight activity is to perform flight tests incorporating propulsion and control subsystems; these are often called control test vehicles. Adding guidance components is a logical next step in what are called guidance test vehicles. Free-flight testing with live ordnance is usually postponed until late in the engineering phase, after confidence is gained in all of the subsystems.

Simulations are used as exploratory and evaluation tools throughout the development, especially during the engineering phase. These begin as purely numerical models and come in many forms, created to serve a variety of purposes. Some are fairly simple and focused; others are quite complex and may support several different areas of interest. As the development effort evolves, some of the numerical submodels in a detailed simulation may be replaced with actual hardware components to form a hybrid arrangement known as hardware in the loop (HWIL). Among other things, this kind of tool can be used to correlate with test results and to artificially "fill in" data between test points.

PRODUCTION-REPRESENTATIVE MODELS

The last phase of the development effort is focused on hardware that matches the intended production configuration. These are called preproduction prototypes, production-representative models, or some similar term. All prior problems should have been worked out and all fixes incorporated. In theory, at least, the design should now be mature and the configuration stable. Virtually all of the hardware produced during this final development phase will be allocated to qualification testing, described in the next chapter.

Milestone Reviews

Military hardware acquisition programs undergo several mandatory reviews as they proceed. As noted earlier, a strike weapon development effort is begun only after a thorough review and approval by high-ranking officials. Additional reviews occur during the course of the development and qualification efforts. The timing and criteria for each review is tailored to the individual program, but the basic intent is to ensure that all aspects of the program are independently examined periodically. Approval to continue is contingent upon passing such reviews. These are therefore major events in the life of a program.

Some Development Observations

It is important to recognize that there are no unimportant parts in the system. The failure of a simple component can lead to a subsystem failure, which can lead to system failure. It's the old story: For want of a nail the shoe was lost; for want of a shoe the horse was lost; for want of a horse . . . ; and so on. A faulty ten-cent electronic component can cause a very expensive strike weapon to fail. Therefore, everything in the system, including existing items, must be carefully scrutinized.

Another reality in development programs is that things seldom go exactly as planned. Some initial ideas don't work out, design or integration conflicts emerge, parts suppliers go out of business or quit supporting particular items, and so on. The conventional engineering development approach is commonly summarized as build, test, analyze, fix, and retest. No amount of theory replaces actual test experience, and from that experience come improved designs. A methodical development program starts with simple pieces and works up to the more complex assemblies, and the test matrix likewise reflects a simple-to-complex, easy-to-difficult theme. This usually uncovers basic design issues at an early stage, where they can be dealt with most easily.

It also needs to be recognized that test failures during development are not necessarily a bad thing. If the failures stem from the application of some new technology or some new requirement, they may well be the result of pushing into an area with unknown design boundaries. Early test failures will help define those boundaries, leading to a better design. Or, as many engineers would put it, you really don't know the limits of something until you break it.

Finally, the only reason that a development program exists is to do something that hasn't been done before. Purchasing more of an existing item does not require a development program. It is only when something new is required that development, under any name, must occur. The extent of development efforts can vary considerably, reflecting the individual needs of the particular programs, but the bottom line is this: if it were simple, development wouldn't be needed.

SEVENTEEN

Qualification

Before a strike weapon is considered ready for production it must be "qualified" for operational use. That is, there must be some kind of verification that the item can withstand the rigors of the operational environment and still function properly. The qualification process involves considerable testing and is the modern equivalent of the old proving-ground activity. It covers mostly performance issues and is an extension of the testing done during development.

Test Plan

Analysis of the experience gained during the development process is often used as a starting point for qualification, but additional testing is needed to verify methodically that the system is ready for use. To produce meaningful results, qualification (or qual) testing should make use of production-representative hardware. Every issue evaluated during qual testing should have been previously addressed during development, but perhaps with slightly different hardware.

Each major program develops its own test and evaluation master plan (TEMP), which is a comprehensive document that encompasses multiple program phases. A well-prepared plan and test data file clearly shows the buildup of test experience over time, indicating what kind of hardware has been subjected to what type of environment. As the system proceeds through development, the test activity progresses from components to subsystems to system. This history helps identify problems and fixes along with achieved performance levels.

The types of qualification tests to be performed and the specific test conditions will be found in the qual test matrix. Some qual testing does not involve actual flights of the weapon. For example, critical components and subsystems may be subjected to severe environmental conditions to verify their continued functionality. In other cases reliability testing may be performed on statistically significant numbers of selected items to verify that they will indeed hold up and function when needed.

The qualification flight test plan is much like an outline for a final exam. It is impractical and unaffordable to check each and every combination of operational conditions during qual testing, just as it is impractical to ask every potential question on a subject during a final exam. Instead, a representative sampling of test points is selected, encompassing a variety of important aspects.

Test Responsibilities

When the bulk of the hardware development activity has been the responsibility of a contractor, which is often the case, much of the development test activity will be conducted by that contractor or its suppliers. Development flight test activity will generally involve the use of government facilities and ranges, but the contractor usually retains responsibility for both the hardware and the basic test approach during that phase of the program. The government provides the launch platforms, test facilities, and support.

Relationships change when it is time to verify system readiness for use. To ensure strict objectivity during the qualification effort, the government takes a more dominant role in both the planning and the conduct of the testing. Put in terms of an automotive analogy, the "customer" (i.e., the government) is evaluating a possible new car, and first decides what the test-drive will include and then does the driving. The "dealer" (i.e., the contractor) simply provides the vehicle, information on its design and history, and support.

It is common to find qual testing handled as two sequential parts: a technical evaluation (TECHEVAL), followed by an operational evaluation (OPEVAL). The first part considers a wide range of issues and is a mix of component, subsystem, and system-level testing, including flight tests. It is generally conducted at the direction of the government program staff.

The second part is a set of mostly flight tests conducted by an independent agency of the military service that is acquiring the new system. This agency represents the end users of the system—the uniformed service members who will ultimately use the new item. As newcomers to the project, agency personnel must first be trained in the handling and use of the system; this in itself is a test of the process, determining whether the training material is adequate for wider application. Then the agency conducts a series of flight tests to assess system readiness for operational use. This phase is truly the final exam for the system.

Dealing with Deficiencies

During the development phase it is relatively common to experience problems in testing. These are carefully examined, and modifications are incorporated into the design to eliminate the deficiency. As the program matures, there should be fewer and fewer such test problems.

The goal is to complete the qualification phase with a perfect test record: no system deficiencies. However, unexpected issues do sometimes arise. If those problems are minor in nature, the system's development might be allowed to continue with only a brief pause for fixes to be incorporated. Problems of a more serious nature are cause for the program to go back into an extension of development. Once the problems have been resolved, some level of requalification will be necessary before deployment is authorized.

On rare occasion, qualification has uncovered deficiencies that were so serious that there was no way that the program could recover in a timely or affordable manner. This has resulted in the program being canceled.

EIGHTEEN

Production

Limited quantities of hardware are indeed produced during the development phase of the program. These early models often include evolving design features as the effort progresses, leading to a series of slightly different hardware and software configurations. As the design matures, the configuration tends to stabilize, such that the qualification hardware represents the design intended for subsequent production.

A decision to proceed into production is a major milestone. It should be based on successful completion of the design and qualification efforts, and is further based on a continued need for the weapon. That latter point may seem a bit surprising to some readers, but it signifies recognition that the operational world may have changed in the time it takes to go from the initial definition of the requirement through development and qualification. A noteworthy example of that kind of external change would be the collapse of the former Soviet Union. When there are significant changes in the perceived threat, military requirements are subject to review and revision.

The production phase of a weapon acquisition program typically represents a much-larger outlay of financial resources than the development and predevelopment phases. The production plan usually involves a multiyear time frame, which means that the government must consider program funding needs over a number of federal budget cycles. Figure 18.1 is a generic look at a representative, multiyear production phase. Competition for resources and the influence of a fluctuating political climate all complicate the acquisition process.

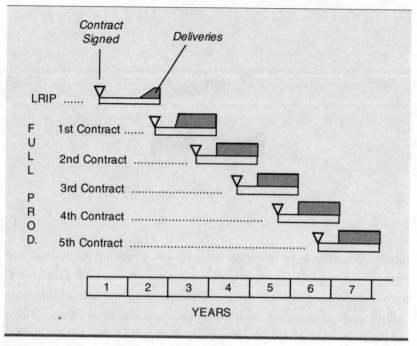

Figure 18.1 Notional Production Contract Sequence

Production Strategy

When substantial quantities of an item are to be produced, the government typically requires that the developer of a weapon, whether that be a contractor or a government team, prepare a production documentation package that is suitable for competitive procurement of the item. The intent is to provide the government with a complete definition of the hardware and software such that any competent supplier could reproduce the item. This allows the government the flexibility to continue with the original developer for the life of the program or to seek competitive bids for production or support of all or major portions of the weapon system.

The typical approach is to begin the production phase with the original developer. If the circumstances appear favorable to the government, later production runs may be opened to competition. In cases involving very high production quantities it would be common to find multiple production contracts let to achieve the desired delivery rates. There are also cases where two contractors bid for production, with contracts let to both for varying percentages of the total run, depending on the price, schedule, and terms offered by the contractors.

The production strategy varies from program to program, and indeed from time period to time period within a long-standing program. It is very much a business-oriented matter, aimed at acquiring the needed hardware in the most advantageous manner, considering quality, cost, delivery schedule, and terms.

Initial Production

The government rarely commits to a complete production run of a new weapon with the first production contract. The more common (and prudent) approach is to start small and build up from there after quality and production capabilities have been demonstrated. The first production contract, for what is often called the low-rate initial production (LRIP) phase, is therefore a first step in the process and involves a limited number of items delivered on a somewhat-relaxed schedule. These first production items may be subjected to more rigorous or widespread quality-control inspections and testing than later ones, just to verify that all is in accordance with the production drawings and specifications.

Service Introduction

Hardware from the first production contract is typically the first wave of hardware delivered to operational units. More will be said in the next chapter about this first operational phase of the program. In terms of the production activity, successful operational introduction of the hardware is often a critical factor in determining whether the program will proceed into full production, or whether it will temporarily remain in the limited-production mode while deficiencies are corrected.

Full Production

Presuming a favorable service introduction, the program usually moves into the full-production phase, also known as full-rate production (FRP). The usual practice is to break the full-production quantity and time frame into several sequential contracts in accordance with the program's production strategy. The production quantities in this phase are much larger and the delivery schedules tighter than those found in the first production contract. Where possible, attempts are made to match delivery rates with the contractor's most economical build rates.

The full-production phase may run just a few years for complex items, such as cruise missiles, that are not used in large quantities, or it may continue for a great many years for less complex items, such as bombs, that are used heavily and consistently.

As expendable items strike weapons tend to be produced and then stock-piled for use when and if needed. Determining the "required" total quantity to be produced is often a matter of serious analysis and debate. Much of the determination is based on best estimates of the likely conditions in future conflicts. Given the potential variability in such estimates, it is not surprising that there can be significant controversy over weapon production quantities. The essential point from an operational perspective is that there should be an adequate mix of weapons available to deal successfully with a wide range of conflict levels.

NINETEEN

Deployment and Support

Output from the production phase of the program feeds the operational needs of the using service(s). When we say that hardware is "deployed" we mean that it is in the hands of uniformed service personnel, ready for use when called upon. Widespread deployment to a variety of operational units requires considerable infrastructure to accommodate needs for such things as storage, transportation, training, support equipment, maintenance, explosive ordnance disposal, and technical assistance. Using an automotive analogy, it is much like introducing a brand-new commercial vehicle model, requiring a new supply-and-support chain from the factory through the distributors and dealerships, which then interface with the businesses that purchase the vehicles and the drivers who operate them.

Initial Deployment

The first deployment of a weapon to an operational unit is usually marked as a program milestone called initial operational capability (IOC). There may only be a few weapons delivered at this point, but this represents the first opportunity that the operators have to actually use the item in either operational training or combat. The initial deployment phase is typically a very busy period during which extensive training is under way while supply lines are established and checked. It is common to experience operational awkwardness and difficulties during this period, leading to considerable effort to work out what seem to be inevitable early glitches in the support system.

Training teams and materials are sent out to key locations in advance of the arrival of the new equipment to provide instruction to handling and operational personnel on system features, field requirements, and functional steps. Technical representatives from the producer or developer of the new weapon are generally on hand during the initial deployment phase to provide direct support as well as communication links back to the development and production organizations. The goal is to make the introduction of the new system as smooth and trouble-free as is practical, and to ensure that support is immediately available to deal with unexpected issues in a prompt and professional manner.

The training and initial-support teams move on to other units as additional hardware is shipped to new bases. The gradual buildup is generally accompanied by initial live firings of the new weapon, either in peacetime training or in combat, which leads to initial evaluations of the overall performance of the new system from the users' perspective. Presuming that the weapon performs well during this phase, the program generally pushes on into full production and expanded deployment.

Full Deployment

If the weapon is intended for widespread deployment, there may be an extended buildup period as sufficient hardware is produced and delivered to a large number of operational units. Each program will have its own, tailored definition of what is considered full deployment, more commonly called full operational capability (FOC). This is the point when a substantial number of service units have a sufficient quantity of hardware and have been adequately trained in its use to constitute the desired military capability. During peacetime most of the weapon deliveries made after that point go to storage depots at strategic locations. Operating units then order weapons from the depots on an as-required basis.

System training continues throughout the operational life of the program. New operational units, or new personnel arriving in existing operational units, receive initial training in weapon handling and use. Refresher training is offered to units that have had the weapon for some time, especially if they are preparing for forward basing overseas. Such activity may include training launches of live or inert weapons on military ranges.

System support activities evolve with time. Initially there is an emphasis on seeking out and correcting deficiencies that may have gone undetected during the development and qualification phases. Later the support functions focus more on regular inspection, service, and maintenance of the weapon. Careful record keeping allows program personnel to observe any trends that may warrant special attention; recurring or chronic issues may lead to sizable corrective efforts.

Weapons may be in service for several decades. Most are designed for long periods of storage without deterioration, but eventually it becomes necessary or prudent to refurbish the dormant stockpile. Such activity may be as simple as conducting a thorough inspection and replacing any faulty components or parts that show reduced performance with age (e.g., seals, rubber items, etc.). In other cases a complete overhaul may be conducted for the purpose of giving the weapon the equivalent of a second life. These more extensive activities may be called service life extension programs, or SLEPs. They are not intended to improve or alter system performance, but simply prolong it.

TWENTY

Upgrades

Over time tactical circumstances evolve that may suggest the need for changes in strike weapon capabilities. Similarly, technical or fabrication advances may offer opportunities for expanded, improved, or less expensive capabilities for system components. Both cases often lead to a review of possible upgrades to existing strike weaponry to achieve the desired results, rather than initiating a new development program.

Minor Upgrades

It is common to observe a pattern of gradual change in strike weapons over their operational lifetime. In many instances this change is relatively minor with regard to the functionality of the equipment; such changes may be transparent to the user, noticeable mostly to the people who maintain and service the equipment. Better internal seals, more robust attachments, more reliable electronic components, a software patch, or a reformulated rocket motor propellant would all fit into this category. These are all important upgrades, but they do not alter the operational envelope of the weapon, nor do they change the launch sequence, etc.

Some minor upgrades are hardly publicized within the operational community. The upgraded components are substituted for their earlier counterparts and carry a different model or serial number, but the overall weapon designation remains the same.

On the other hand, some minor upgrades are sufficiently extensive that they warrant an overall weapon nomenclature change. If, for example, the first fielded version of the weapon were listed as AGM-345A, the upgraded version might be identified as AGM-345B.

Major Upgrades

A change in nomenclature is a certainty with a major upgrade. This type of upgrade will generally produce functional- or operational-envelope changes in the weapon, which are sometimes very substantial. It is common to hear of weapon "variants" when discussing upgrades of this nature, which alter or expand the role or level of capability of the weapon.

The Maverick air-to-surface missile serves as a convenient example of a weapon that evolved significantly over time. It began as a U.S. Air Force rocket-propelled antiarmor weapon with a conical-shaped charge (CSC) for its warhead and a television (TV) seeker. The AGM-65A and AGM-65B provided very accurate weapons for use during day, clear line-of-sight conditions. But there were occasions when troops on the ground could observe a target that was not easily visible to the pilot in the delivery aircraft. To solve that issue, a laser seeker was fitted to the Maverick in place of the TV seeker, which allowed the missile to home in on the energy reflected from the target that had been designated by the ground troops using a laser. Thus the AGM-65C Laser Maverick was born.

The desire for a direct-fire weapon usable at night led to the development of an imaging infrared (IIR) seeker for the AGM-65D Maverick. Then the U.S. Marine Corps determined that its needs for a direct-fire weapon could be met if the Laser Maverick had a heavier and more robust penetrating-blast (PB) warhead, along with a low-smoke rocket motor. This led to the AGM-65E.

The U.S. Navy found that its needs for a direct-fire weapon could be met with something similar to the Marine Corps version but with a modified IIR seeker suitable for attacking ships as well as land targets. That became the AGM-65F. Meanwhile, the U.S. Air Force found that it could find good use for a modified Navy Maverick, and thus launched the AGM-65G.

All of these variants of Maverick have the same external dimensions and configuration. The later models are heavier, due to the greater weight of their warheads, but it is difficult to tell them apart from a distance. However, there are notable functional differences in their roles and capabilities.

Changing roles are also found in the AGM-84 Harpoon missile family. The mission of the baseline Harpoon is ship attack. Several versions of Harpoon have evolved that mission focus, each primarily reflecting upgrades to the all-weather-guidance subsystem.

However, there are also Harpoon evolutions that altered its combat role. The first was the AGM-84E SLAM (standoff land attack missile), which substituted a modified Maverick IIR seeker for the baseline radar seeker and added a data link. This allowed the missile to be used against targets other than ships. Later came the AGM-84H SLAM-ER (SLAM–expanded response), which incorporated upgraded guidance components in a repackaged and reconfigured nose along with pop-out planar wings for better aerodynamic performance.

These are just two of the numerous examples of upgrades made to weapons as needs or opportunities appear over the long life of most systems. In most cases upgrades or variants were not a preplanned aspect of a development program. The initial focus was entirely on meeting the stated requirements. Subsequent upgrades came about where practical and affordable, but not as a result of initial planning. However, that historic trend may be altered by more recent developments, where initial planning includes an eye toward later change.

The AGM-154 Joint Standoff Weapon (JSOW) may serve as an example of a development that anticipated evolution. From the very outset the program documents openly described a weapons family, with three initial variants, all of them glide weapons. A fourth major variant, powered by a small turbojet, has also been pursued, and heavier versions with larger payloads have been discussed. Time will tell whether additional versions spring forth.

TWENTY-ONE

Retirement

Ultimately military hardware reaches the end of its useful life. It may be a case of excess maintenance required to keep it operational, the lack of replacement parts, or simply that its capabilities no longer meet minimum needs. Whatever the reason, the time will come when a system is declared to be at the end of service life (ESL). When that occurs, the equipment is retired from service and removed from operational stockpiles.

Phaseout

Equipment retirement is seldom an abrupt event. There generally is a gradual phaseout of the old, sometimes accompanied by the phase-in of something new. The retirement of a strike weapon involves multiple locations and organizations, everything from operational units to support and training facilities, storage and maintenance depots, even parts suppliers. A well-run retirement requires considerable planning to remove affected equipment and data in an orderly manner without disrupting essential services or the flow of operational matériel.

In many respects the mechanics of phaseout is much like initial deployment, only in reverse. Weapons and related items are withdrawn from operational units and returned to intermediate locations, where the matériel is collected and consolidated. Further withdrawal to final installations within the continental U.S. (CONUS) then occurs, sometimes as a slow logistics stream using transportation on a noninterference basis. With time, none of the hardware or unique items of support equipment or data remains in the normal logistics pipeline.

Demilitarization

Strike weapons that have been retired from service pose something of a problem for the government. To begin with, they contain high explosives. For that reason alone, they can't just be piled up in some unsecured scrapyard. Second, they may contain toxic materials or other items that are inherently dangerous (such as pyrotechnics or explosively actuated devices). Finally, even though they are "old," they may contain elements that are still considered to be classified. Because of these safety and security issues, some kind of demilitarization procedure will be used to render the items safe and nonsensitive before any general salvage or scrap activity begins. Thoughtful planning is needed to ensure that the process deals with all such issues.

The process usually begins with the separation of major parts. This allows the warhead and rocket motor components, if so equipped, to be diverted to appropriate additional processing areas. Other energetic devices also are removed and separated from the remainder, which can then be handled with fewer restrictions.

Economics will usually dictate how the separated parts will be processed from this point onward. Rocket motors are sometimes stockpiled for possible use in unrelated test operations. Warheads are usually either destroyed or disarmed by having the explosive removed; in the latter circumstance the empty warhead case may then be suitable for metal salvage. Body parts are often salvaged for metal content.

Other components may be destroyed if classification is an issue, stockpiled if there is some perceived future use, or salvaged for whatever residual value might be available. Salvage can become complicated whenever toxic materials are present.

Records and Data Retention

Disposal of weapon hardware basically eliminates the physical presence of the system, except for a very limited number of examples that might remain in museums or as display items on military bases. However, a paper trail usually remains to some degree. The scope of documentation remaining in federal records repositories varies widely, and the cataloging of such data is likewise variable. Nevertheless, there is an ongoing practice of government retention of at least some records of every strike weapon system developed in this country. Access to such records may remain restricted, usually due to original classification markings remaining on documents because of the limited availability of authorized reviewers who could downgrade the classification after the system was retired.

Epilogue

The end of the book seems an appropriate place to reflect on some reminders, some observations, and a few thoughts about the future. In keeping with the tone of the rest of the text, these will be addressed at a summary level and without excess elaboration.

Perspective of the Text

To a large extent what you find in this book is a historical perspective, looking back on both operational and development practices that have been in use during the past half century or so. The evolution of strike warfare has been described up to a point in time. Acquisition practices are likewise described in the past tense. Readers should therefore be cautious about assuming that the future will mirror the past. On the other hand, an understanding of the fundamentals that guided prior strike operations and strike weapon developments in the past will definitely provide the reader a better foundation for understanding what occurs in the future.

A second important reminder deals with the depth of the subject matter included in this book. With few exceptions, the discussion has simply provided an overview of conventional or "textbook" situations. This was done quite intentionally to avoid further complicating a subject that is inherently complex. Once again, the focus was on basics rather than on exotics. Readers will want to be aware that there are numerous examples of successful *non*standard practices in both operations and developments.

Observations

Without exaggeration, thousands of pages have been written on the subject of lessons learned from prior operations or developments, many of them painful. Many of the operational lessons remain classified and cannot be discussed here. However, one of the most basic lessons that apparently requires frequent repeating in the operational world is best expressed in the following sentence: *Never assume that your opponent is stupid.* Operational arrogance is foolish at best and can lead to deadly consequences. One well-aimed (or lucky) 7.62-mm bullet can take down a multimillion-dollar flying machine, and a dedicated insurgent with a minor bit of training can deliver a powerful explosive device.

THE ACQUISITION SYSTEM

In the development world, there have been numerous studies conducted to evaluate acquisition policy and practices in hopes of finding better ways to conduct business. Each time this has been done, a list was produced of findings and recommendations. Adjustments were made, new instructions were issued, terminology was changed, and we proceeded onward. It has been hoped that significant improvements would come out of this process, but it generally hasn't taken long before new complaints surfaced about faults in the acquisition system.

Part of the problem in the acquisition system is the migratory nature of the participants. This is especially true in the upper levels of both the military and civilian decision makers and managers, where individuals tend to remain in position on a given program for only a fraction of its life. As administrations change, political appointees change, as do underlying policies. Military officers rotate through assignments on a regular basis, and senior civilian staff members move on to new assignments as career advancements become available. The result can be a loss of personnel stability in the oversight and management of lengthy programs, which can introduce uncertainty or perturbations in the acquisition process.

Another acquisition issue of human origin is the manner in which programs are funded. The program manager outlines a multiyear effort to develop, qualify, and produce the item in question. Checks and balances are written into the plan, identifying major program reviews at key points to ensure that everything is going well before proceeding into subsequent phases of the effort.

Contracts are negotiated and signed presuming successful activity according to the program plan. Everything is neatly laid out and agreed to, there are no overwhelming technical issues in the way of success, and the program appears rock solid . . . until the congressional budgeting process comes into play.

The problem with a multiyear military acquisition program is that federal funding occurs on an annual basis, with no guarantees that the budgeted amount will actually be provided. Not only can the annual allocations change, Congress can attach wording to alter certain aspects of the budget request. Furthermore, during the life of most large acquisition programs there will be at least one presidential election, leading to the possibility of changes in defense strategy, priorities, or policies while the program is under way. Do politics intrude into the acquisition world? You bet they do.

Shifting our focus now to a less volatile matter, there are very strict rules for the awarding of contracts in military acquisition programs. A formal process has been emplaced to ensure a fair and unbiased selection that is based solely on merit and value. Abiding by this rigid structure means that extra time and effort is required to evaluate contractor proposals and make an award, as compared with routine commercial practices. The regulations include a requirement that the government personnel involved in the proposal evaluation and contract award process avoid any actual or perceived conflicts of interest with parties involved in the potential contract.

This structured "source selection" process does add a burden to the weapons acquisition process. There is no doubt that time and funding could be saved by abandoning it, as some have suggested. On the other hand, there have been enough examples of abuses even with the current system in place that it would seem foolish to simply eliminate all the constraints. Perhaps the better lesson to be drawn from this arena is that human beings will be tempted to break the rules whenever financial gain is a possibility, but a reasonable reduction of the source selection structure, coupled with reasonable oversight, would serve the taxpayers well.

Legislators and decision makers have also created a very lengthy and complex set of federal acquisition regulations in general. Many of the provisions appear to have been enacted to prevent past errors and abuses, which makes it difficult to argue with the rationale behind their adoption. Unfortunately the sheer magnitude of these regulations and associated compliance measures add a very large administrative burden to weapons acquisition programs. When you couple this additional responsibility with challenging performance and operating-environment requirements, you end up with expensive hardware.

TECHNICAL MATTERS

At the onset of a program, and indeed whenever program goals are considered, there is a temptation to add just a little bit more performance or a few more features to the list of requirements. It is probably a basic human trait to want our creations to excel, to be more than just OK. In the weapon development world, this tendency to replace "good enough" with "better" is known as "requirements creep."

When dealing with expendable items such as strike weapons, it is very important that the development team be given a set of requirements that is realistic, essential, and stable. Capabilities that would be "nice to have" or "optional" are, by definition, not essential, but they do add complexity and cost. Expanding basic requirements while writing the system spec or the contract will have the same detrimental result as altering the basic requirements document. It's a hard lesson to accept for overachievers, but it has plagued many a program.

Another requirements/specification issue deals with the operating environment. In the exuberance of initial goal setting, it is tempting for planners to require that military hardware be fully functional at any location on Earth, during any season and weather, day and night. This "anytime, anywhere" approach can lead to environmental and performance requirements that can only be satisfied with very expensive solutions. It is better to be realistic about operating conditions and end up with an affordable item that is usable in most places, most of the time.

On more of a development process note, the increased application of digital computing technology in weapons has brought with it increased utilization of specialized software. As this happens it becomes more and more important that software development, verification, and qualification be addressed with just as much attention and rigor as hardware development, verification, and qualification. Discovering software quirks during combat is not an acceptable situation.

Finally, as computer tools have become more and more capable, there has been a natural increase in the use of computer simulations during the development and evaluation process. Heavy use of sophisticated simulations has the potential to substantially reduce the amount of actual field testing required to explore weapon performance within the breadth of its operating envelope. However, it is essential that sufficient actual testing be conducted to verify the accuracy of those simulations. Excess reliance on simulations without sufficient actual verification can lead to distorted expectations and inadequate checking of boundary conditions. Again, these are things you would rather discover long before final qualification or deployment.

The Future of Strike

During the Cold War era, the United States focused most of its strike warfare attention on a robust, technically capable opponent: the Soviet Union. In many respects we prepared to deal with an adversary that was often a mirror image of ourselves in terms of military capabilities. We needed to be able to counteract large numbers of sophisticated weapon systems manned by well-trained warriors who wore distinctive uniforms . . . just like ourselves.

The collapse of the Soviet Union and the subsequent emergence of less familiar threats was a major disruption to American defense planning. Preparation for "high-threat" conflict became less an issue than coping with elusive small groups who blended in with the local population while engaging in classic hit-and-run tactics using low-technology weapons. And yet there remained potential regional threats possessing high-technology systems.

If there is anything that the history of human conflict should teach us, it is that opponents seek to find and exploit weaknesses in each other. During the American Revolution, the rigid battlefield formations and strategy of the redcoats was exploited by the "less capable" minutemen, who used flexibility and cover to their advantage. It wasn't that the British approach was wrong, for it was definitely appropriate to contemporary land battles in Europe. The problem for the British was that they were not able to adapt quickly to a different type of conflict.

There is a similar danger in American defense planning. Adapting to a changing world has always been a challenge. We are more skilled at planning for the last fight than for the next one, partly because it is far easier to review recent history than to forecast the future.

There is a temptation to focus on immediate circumstances and needs, and develop a defense strategy that fits those particulars at the expense of other situations. There are even strong political and economic pressures to adopt such an approach because it is single-minded in purpose, addresses clear issues, and doesn't require the expenditure of resources on other possibilities. Unfortunately this is also a very shortsighted and ultimately dangerous approach. It signals potential adversaries that America is putting all its warfighting eggs in a single basket, which then allows them to begin to exploit the weaknesses that will result in our nonsupported areas.

The more prudent approach is to shift priorities and emphasis to adapt to changing circumstances while maintaining a core level of capability in all appropriate areas. For example, if long-range strike is less of a concern in the near term, then it can move down in priority without being completely ignored.

Moving from the philosophical level to more specific issues, we find some interesting trends in recent years that differ significantly from historical practice. Some of these are directly tied to the application of new technologies that simply weren't available in prior times.

Just about every recent strike platform, strike weapon, or strike weapon upgrade has made use of GPS navigation in some form. The worldwide availability of GPS signals has, without exaggeration, revolutionized military and civilian navigation. Tiny, lightweight receivers can be installed on just about any military device, providing location or guidance data to whatever display or steering mechanism is available. GPS-guided munitions have proliferated at an astonishing rate, promising excellent delivery accuracy day or night in virtually any kind of weather. The only thing they require is (1) an acceptable launch condition, (2) accurate target coordinates, and (3) GPS signals.

As favorable as this technology may be, GPS does introduce two important cautions. First is the matter of accurate target coordinates. A GPS-guided weapon navigates to whatever geographic "address" it has been provided. If that address is not correct, it doesn't matter to the weapon; it will blindly go to the coordinates that were entered. Targeting accuracy and careful data entry therefore become very critical.

The second operational concern with GPS-guided weapons is their reliance on usable, uncorrupted GPS signals. You can be certain that this reliance has not escaped the attention of potential adversaries. An obvious counter to GPS-guided weapons would be some means of jamming or altering the signals, so as to degrade navigation performance. Backup guidance modes would therefore seem prudent for such an eventuality, lest we find our weapons inventory suddenly unusable in the presence of a capable adversary.

Another important change in American combat operations involves the use of unmanned air vehicles. UAVs rapidly transitioned from experimental devices used in limited numbers to widely used, fully operational systems. They range from very small, man-portable battlefield surveillance drones to large, armed combat aircraft. Some have the ability to remain aloft for extended periods, providing multiple forms of airborne observations to combat theater commanders.

These various unmanned machines offer unprecedented visibility of the battle. Some also serve as proxy warriors, launching air-to-surface weapons at targets observed from their unique vantage points. Having the ability to locate and attack targets remotely does bring up the matter of rules of engagement and combat ethics. At the core of this issue is a simple but troubling question: *What degree of autonomy is acceptable in unmanned combat operations?*

Unmanned observation platforms are not a direct concern, but when offensive action is taken by an unmanned platform one must carefully consider exactly how that action occurs. For example, how was the target identified? What assurance is there that the targeting is correct? How is the weapon launch decision made? Where does the responsibility lie for unintended collateral damage?

When humans are part of the targeting and launch decision process, the UAV serves as a remotely operated warrior, obeying human commands. If, on the other hand, decision making resides within the UAV with no human review or control being applied, we have an entirely different situation: a truly autonomous unmanned combat air vehicle. This is no longer science fiction.

Yet another major change in combat operations involves the rapid expansion of electronic communications linkages between participants. From the foot soldier to military satellites, various types of communication links pass information back and forth. The amount of data flowing around the combat theater has increased exponentially, to the point that there are serious concerns about overwhelming the human capacity to absorb and make use of the information.

The advantage of this expanded communications capability is that field commanders can obtain far more information about their surroundings than ever before. When everything is working well, they can track the movement of their forces with great precision; and, depending on the accuracy of the intelligence and surveillance data, they have a better awareness of their opponents' situation and movements. Better awareness generally means increased combat effectiveness.

The downside of this intensive communications activity is that it could potentially be exploited by a capable adversary, either through interception for intelligence purposes, corruption or spoofing to introduce errors, or denial (e.g., jamming). Becoming too reliant on rapid and uninterrupted data flow could become an Achilles' heel. Backup modes of operation are therefore important to ensure some reasonable level of combat effectiveness in the presence of electronic countermeasures.

Nontraditional warheads have emerged as another evolutionary change in strike warfare. Something like a burst of intense electromagnetic energy to disable or destroy nearby electronics offers a different approach to neutralizing certain targets. Such a device should also minimize loss of life and structural damage. It does, however, introduce a damage assessment challenge: How can you be certain that the targeted electronics have indeed been neutralized?

Finally, the advent of increasingly more accurate guidance modes has allowed many (but not all) strike operations to be conducted with smaller weapons. This is basically a case of placing the weapon exactly where you want it, which then means that a smaller warhead can destroy the target. Smaller weapons are also advantageous for advanced aircraft, where signature control argues for internal carriage in small weapons bays. Thus we see a trend toward small, precision guided weapons.

Much of the surface target array can indeed be attacked successfully with well-delivered small weapons. However, there are some robust targets that are difficult to destroy with small warheads, no matter how accurately delivered. As a result there is a need to ensure that an adequate inventory of larger weapons is maintained to address those more challenging targets.

Taken in sum, these trends suggest that future strike warfare will likely involve increased application of precision strikes against specific target elements, with less collateral damage and fewer weapons expended . . . when the operating environment permits. This broad projection is only applicable when friendly forces dominate the electronic battlespace and have subdued high-threat air defense elements. When those conditions cannot be met, more historic strike tactics are likely to be required.

Appendix A

Glossary of Terms and Abbreviations

AAA	antiaircraft artillery; guns used as a defense against aircraft
AAM	air-to-air missile
AARGM	advanced antiradiation guided missile; upgraded AGM-88 HARM missile
AAW	antiair warfare; operations targeting enemy aircraft
AEW	airborne early warning; airborne radars and other sensors (e.g., AWACS)
AFV	armored fighting vehicle
AGM-	air-to-ground missile; U.S. military designator
ALCM	air-launched cruise missile; AGM-86
AoA	analysis of alternatives
APAM	antipersonnel-antimatériel
APC	armored personnel carrier
ARM	antiradiation missile; a missile that homes in on radar antennas
ASM	air-to-surface missile
ASW	antisubmarine warfare; operations targeting enemy submarines
ASuW	antisurface warfare
ATO	air tasking order
autopilot	A device that maintains flight along a desired trajectory.
AWACS	airborne warning and control system; i.e., the E-3 aircraft
ballistic	A mode of flight that does not intentionally generate lift.
BDA	bomb damage assessment; i.e., how much damage was done?
BGM-	multiple-platform strike weapon; U.S. military designator
BIT	built-in test

BLU-	DoD designator for a bomb or submunition
bomblet	A small munition; groups (clusters) of bomblets are carried in dispensers.
C3	command, control, and communications
C-ALCM	conventional air-launched cruise missile; i.e., the nonnuclear AGM-86C
CAP	combat air patrol; fighter aircraft used to create a protective screen
CAS	close air support; aircraft operations in support of ground troops
cat	A catapult takeoff from an aircraft carrier.
CATM	captive air-training missile
CBU-	cluster bomb unit; U.S. military designator for a bomblet
CEB	combined-effects bomblet; i.e., the BLU-97 submunition
CEM	combined-effects munition; i.e., the CBU-87 dispenser with submunitions
CI	close-in; short-range
CNWDI	Critical Nuclear Weapons Design Information (security term)
COEA	cost and operational effectiveness analysis; superseded by AoA
CONUS	continental United States
COTS	commercial off the shelf; equipment available commercially
CSAR	combat search and rescue
CSC	conical-shaped charge; a type of warhead
CV	aircraft carrier; U.S. military designator
DL	data link
DoD	Department of Defense
ERDL	extended range, data link
ESL	end of service life
EW	electronic warfare; operations using electronic sensors, jammers, and devices
FAC	forward air controller; air or ground operator who designates targets
FAE	fuel-air explosive
FBM	fleet ballistic missile
FC	fire control
FEZ	fighter engagement zone
fin	A fixed stabilizing surface or a movable control surface.
FLIR	forward-looking infrared; an imaging infrared sensor
FOC	full operational capability

frag	fragmentation
FRP	full-rate production
fuze	A triggering device used to detonate a warhead.
GAM	GPS-aided munition; i.e., the GBU-37
GBU-	guided-bomb unit; U.S. military designator
GCI	ground-controlled intercept
GFE	government-furnished equipment
GFI	government-furnished information
GFM	government-furnished matériel
GFP	government-furnished property
GO	OK to proceed, as opposed to NO-GO
GP	general-purpose
GPS	Global Positioning System; satellite navigation system
guidance section	That portion of the weapon that contains the seeker, computer, autopilot, etc.
HARM	high-speed antiradiation missile; i.e., the AGM-88
HE	high explosive
HPM	high-power microwave
HWIL	hardware in the loop; hybrid simulation technique
IADS	integrated air defense system
ICBM	intercontinental ballistic missile
IIR	imaging infrared
INS	inertial navigation system
IOC	initial operational capability
IR	infrared
IRBM	intermediate range ballistic missile
ISR	intelligence, surveillance, and reconnaissance
JASSM	Joint Air-to-Surface Standoff Missile; i.e., the AGM-158
JDAM	Joint Direct Attack Munition; a GPS-guided bomb family
JSOW	Joint Standoff Weapon; i.e., the AGM-154 family
LGB	laser-guided bomb
LLLGB	low-level laser-guided bomb
LOAL	lock on after launch
LOBL	lock on before launch
LOC	line(s) of communication
LRIP	low-rate initial production
LSC	linear-shaped charge; a type of warhead
M	munition; DoD designator

Mach	The ratio of velocity to the speed of sound; Mach 0.5 is half the speed of sound.
MANPADS	man-portable air defense system; shoulder-fired SAM
MEZ	missile engagement zone
MIL-SPEC	military specification
MIL-STD	military standard
Mk	Mark; U.S. Navy designator
mm	millimeter
mmw	millimeter wave
MOAB	massive ordnance air blast; i.e., the GBU-43
Mod	Modification; U.S. Navy designator
MOP	massive ordnance penetrator; i.e., the GBU-57
munition	A device containing explosive or similar energetic material.
NATO	North Atlantic Treaty Organization
NOFORN	no foreign (security marking)
NO-GO	do not proceed; opposite of GO
OPEVAL	operational evaluation
ORCON	originator control (security marking)
ordnance	Generic term for military weapons.
payload	That portion of a missile or weapon that includes the warhead or its nonexplosive substitute.
PB	penetrating-blast (a type of warhead)
Pd	probability of damage
Pk	probability of kill
propulsion section	That portion of the missile containing the engine and fuel.
qual	qualification
R&D	research and development
RDT&E	research, development, test, and evaluation
RFI	request for information
RGM-	ship-launched strike weapon; U.S. military designator
ROE	rule of engagement; also, a set or series of such rules
SAM	surface-to-air missile
SAR	special access required (security term)
SCI	sensitive compartmented information (security term)
SDB	small-diameter bomb
SEAD	suppression of enemy air defenses

seeker	A device that homes in on some target feature, signature, or characteristic.
SLAM	standoff land attack missile; i.e., the AGM-84E
SLAM-ER	standoff land attack missile–expanded response; i.e., the AGM-84H
SLEP	service life extension program
SNTK	special need to know (security term)
SOAD	standoff outside area defenses; long standoff
SOB	Shrike On Board; a surface-launched Shrike missile
SOPD	standoff outside point defenses; medium standoff
SOTD	standoff outside theater defenses; very long standoff
spec	specification
SSM	surface-to-surface missile
submunition	A small munition, groups of which are carried in a dispenser; *see* bomblet.
TECHEVAL	technical evaluation
TEL	transporter-erector-launcher
TEMP	test and evaluation master plan
TLAM	Tomahawk land attack missile; i.e., the nonnuclear BGM-109
TOT	time over target
trap	An arrested landing on board an aircraft carrier.
TSSAM	tri-service standoff attack missile; i.e., the AGM-137
TV	television
UAV	unmanned air vehicle
UCAS	unmanned combat air system
UCAV	unmanned combat air vehicle
UGM-	Underwater (submarine)-launched strike weapon; U.S. military designator
UNREP	underway replenishment
USAF	U.S. Air Force
USMC	U.S. Marine Corps
USN	U.S. Navy
VA	attack aircraft unit; U.S. Navy designator
VERTREP	vertical replenishment
VF	fighter aircraft unit; U.S. Navy designator
VFA	fighter/attack aircraft unit; U.S. Navy designator
VQ	electronic-warfare aircraft unit; U.S. Navy designator

warhead	An explosive and its surrounding case.
warhead section	That portion of the weapon that includes the warhead, fuze, TDD, booster charge, and any ancillary equipment items.
wing	A surface intended to produce lift.
WNINTEL	warning notice—intelligence sources and methods (security term)

Appendix B

Generic Design Criteria

Chapter 15 provided an overview of design criteria for strike weapons. In the interest of brevity, much that could have been said about this important topic was left out. This appendix has therefore been created to add a bit more to the story for those who have additional interest in the subject. As is the case throughout this book, the truly technical has been limited in scope and terminology, leading to simplified treatment of a complex subject.

Specification Provisions

As noted in chapter 15, the system specification (spec) serves as the technical foundation for the development effort. Those who are familiar with such efforts will point out that some parts of the system spec are largely boilerplate text, meaning that the same words are found in many different weapon specifications. However, most of the document will be tailored to the specific situation.

The abbreviated system spec outline found in chapter 15 has been expanded below to illustrate a more complete, representative outline of the document.

1. SCOPE
   1.1 Purpose
2. APPLICABLE DOCUMENTS
   2.1 Government Documents
   2.2 Nongovernment Documents

3. REQUIREMENTS

3.1 System Definition

    3.1.1   General Description

    3.1.2   Mission

    3.1.3   Threat

    3.1.4   System Diagram

    3.1.5   Interface Definition

    3.1.6   Operational and Organizational Concepts

3.2 Characteristics

    3.2.1   Performance

    3.2.2   Physical Characteristics

    3.2.3   Reliability

    3.2.4   Maintainability

    3.2.5   Availability

    3.2.6   System Effectiveness

    3.2.7   Environmental Conditions

3.3 Design and Construction

    3.3.1   Materials, Processes, and Parts

    3.3.2   Workmanship

    3.3.3   Interchangeability

    3.3.4   Safety

    3.3.5   Human Interfaces

3.4 Documentation

    3.4.1   Drawings

    3.4.2   Specifications

    3.4.3   Technical Manuals

    3.4.4   Test Plans

    3.4.5   Configuration Management and Control

3.5 Logistics

    3.5.1   Maintenance

    3.5.2   Supply

    3.5.3   Facilities and Equipment

3.6 Personnel and Training

3.7 Functional Area Characteristics

3.8 Precedence

4. VERIFICATION

    4.1 Verification Method Matrix

    4.2 Examination, Analysis, and Simulation

    4.3 Laboratory Test

    4.4 Qualification Test

    4.5 Technical Evaluation (Field) Test

    4.6 Operational Evaluation Test

5. PACKAGING

6. NOTES

Development teams are made up of people representing different specialty areas who focus on one or more of the specific topics listed in the spec outline. Subteam leaders oversee the efforts on related topics, and one or more system engineers will be responsible for bringing the overall effort together in a well-integrated manner.

Design criteria are found embedded throughout the system spec, which means that the entire document must be familiar to at least the system engineering people and the subteam leaders. It is common practice at the onset of a development effort for these people to carefully review the spec and identify all specific requirements and criteria. From this process key parameters are listed, which may be used to create a matrix of criteria that can then be used at the working level to guide and evaluate development.

The challenge of translating specification language into design criteria can be illustrated with a few examples for a hypothetical air-launched strike weapon being developed for a hypothetical launch aircraft.

PHYSICAL CONSTRAINTS

- *Spec language.* The AGM-345 guided missile shall be physically compatible with internal and external carriage on the F/A-55 aircraft without the need for clearance modifications to that aircraft. In no case shall the overall length of the assembled guided missile exceed 168 inches. Gross weight of the assembled guided missile shall not exceed 1,500 pounds.

- *Criteria.* Weight not more than 1,500 pounds, length not more than 168 inches. Configuration must not intrude into F/A-55 clearance limits during upload, carriage, and launch (requires physical compatibility checks of both internal and external carriage locations).

CARRIAGE

- *Spec language.* The AGM-345 guided missile shall be capable of safe and normal captive carriage operation within the full flight envelope of the F/A-55 aircraft with no restrictions on allowable maneuvering.
- *Criteria.* Missile must withstand the loads, vibration, noise, and thermal environments that exist within the unrestricted flight envelope of the launch aircraft (requires a review of the flight envelope and may require test flights to gather data on those environments).

LAUNCH ENVELOPE

- *Spec language.* The AGM-345 guided missile shall be capable of level launch, safe separation, and smooth transition to controlled flight with 300 feet launch ground clearance (i.e., launch at 300 feet above ground level) within the launch envelope shown in Figure xx.
- *Criteria.* Capable of successful launch down to within 300 feet of the surface, from altitude and speed (or Mach number) combinations illustrated in the referenced figure.

FLIGHT ENVELOPE (STANDOFF)

- *Spec language.* The AGM-345 missile shall be kinematically capable of achieving a maximum on-axis standoff range of no less than 30 nautical miles when launched from a level attitude at an altitude of 30,000 feet above mean sea level at an indicated Mach number of 0.80. For definition purposes, the kinematic requirement assumes that target is stationary at zero feet above mean sea level under Standard Day atmospheric conditions with no wind.
- *Criteria.* Capable of flying a specific distance under well-defined conditions. Standoff under other launch conditions, such as different altitudes and speeds, would differ.

Each topic in the system spec is considered in similar fashion, leading to the combined matrix of requirements for the development team.

Environmental Criteria

One of the most important differences between developing goods for the commercial market and developing military hardware is that the latter must survive and function reliably after exposure to some very difficult environmental conditions. One way to gain an appreciation for those conditions is to follow the life cycle path of a representative strike weapon as it moves from factory to eventual destruction at the target.

An example of the environmental-design flow path is illustrated in figure B.1. The path begins at the top of the figure, where the weapon is manufactured, assembled, and subjected to acceptance testing prior to delivery to the government. The flow then proceeds downward, indicating various possible transportation, handling, and storage modes as the hardware makes its way toward the ultimate user, an operational unit. Further storage and handling occur at the user level, prior to the weapon being selected for an operational mission. Different types of launch platforms result in different types of environments. Finally, the weapon is launched and proceeds to the target, where it is destroyed.

The following sections provide a simplified summary of such a series of environmental exposures.

FACTORY

Once assembled and having passed whatever acceptance testing is required, the weapon goes into predelivery storage at the factory. Simple munitions such as bombs may be handled and stored in a "bare" condition, while most other items are placed in protective containers. Complex weapons such as guided missiles are generally delivered as assembled items, with perhaps wings and fins packaged separately. Other weapons may be delivered as separated major components. A GPS-guided bomb, for example, might be assembled at the user level from a bomb, a separate fuze, and a separate guidance kit that includes the tail cone and fins. Each of those pieces is likely to be delivered from a different manufacturing source.

Predelivery storage at the factory is generally a rather short period (i.e., a few weeks at most) in a relatively benign environment, and is not discussed here.

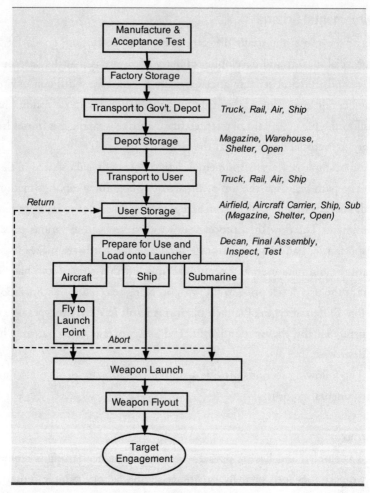

Figure B.1 Simplified Environmental-Exposure Flow Path

TRANSPORT TO DEPOT

When weapons hardware is delivered to the government, it generally goes initially to some kind of military depot, which is a facility where munitions and related hardware can be safely stored, handled, and prepared for shipment to users. Getting hardware from the factory to the depot may involve one or more transportation modes, including overland travel by truck or rail, water-borne transit by ship or barge, or even cargo aircraft delivery. Bombs and other simple hardware items may be shipped in an exposed condition on pallets, while other items will be inside containers.

A variety of environmental exposures must be considered for transportation and associated handling. These include:

- Handling (i.e., movement by hand; using forklifts, dollies, cranes, etc.): shock; vibration; accidental drops; rapid temperature change (thermal shock); dust; accidental fluid contamination; exposure to rain, snow, or decontamination washdown
- Transportation (i.e., shipment by various land, sea, and air methods): shock; vibration; accidental drops; temperature and humidity extremes; pressure (altitude) changes; exposure to dust, salt (seawater) spray, and precipitation (rain, snow, hail)

DEPOT STORAGE

The first stop after manufacture is typically at a military storage facility (depot) within the continental United States. Explosive and propulsive items are typically stored in sturdy earth-backed concrete enclosures called magazines (sometimes referred to as igloos). For safety, the quantity of energetic devices stored in any given magazine is limited, and the magazines are separated from each other (and from other parts of the facility) to reduce the spread of damage in case of an accidental explosion. During periods when there is a surge in the production of weapons, magazine capacity may be exceeded, leading to temporary storage of hazardous items in separated stacks out in the open or under simple roofing that offers only rain and sun protection. Munition components that are not hazardous are generally stored in more conventional, warehouselike surroundings.

Items may be exposed to the depot storage environment for many years before being forwarded to a combat unit for use. Some units may be transferred to other depots to replenish stocks that have been depleted there. Others may be returned to the factory or a rework facility for upgrades, modifications, or repairs before being sent to a combat unit. And some may never leave depot storage before the weapon is withdrawn and retired from service use.

Since it is unknown which items may be in very-long-term storage, or what kind of storage condition may apply (i.e., magazine, covered shelter, unprotected), weapons must be designed to withstand all of the reasonably possible storage conditions and combinations.

Transport to User

The handling and transportation environment for weapons destined for combat units is basically similar to the factory-to-depot environment previously discussed, but some weapons may be exposed to additional environmental conditions. Ship- and submarine-launched strike weapons will experience some dockside handling as they are loaded on board the vessels. Weapons that are delivered on board aircraft carriers must also be designed to tolerate the dockside environment as well as a unique set of conditions associated with at-sea replenishment of weapons, which employs either a cable transfer mode between moving vessels at sea or a helicopter delivery mode with the weapons suspended beneath the helicopter. These two modes are referred to as underway replenishment (UNREP) and vertical replenishment (VERTREP) respectively.

User Storage

Storage conditions at the user level vary quite significantly, simply because the settings are so diverse. User storage may occur at basing alternatives such as air bases, submarines, ships, and aircraft carriers, and those bases (stationary or floating) may exist in diverse locations around the world. Overall, the potential user storage environment includes everything from Arctic to desert to tropical settings. The particular mission and launch platform category will dictate which of these alternatives apply to a specific strike weapon, with user storage environmental conditions tailored to that application.

Preparation for Use

The separate pathway for different basing alternatives that began with user storage continues as weapons are prepared for launch and then expended.

Ship-Launched Weapon

Most current U.S. strike weapons that are launched from surface ships are stored within an enclosed launcher and do not require physical preparation prior to use. An electronic test may be performed to verify readiness for use. Mission data (i.e., instructions for postlaunch activity) are loaded into the weapon and usually verified before the strike weapon is considered ready for launch.

Submarine-Launched Weapon

Strike weapons that are launched from U.S. submarines may be stored, ready for use, within vertical launcher tubes, or they may be stored horizontally in the torpedo room, ready for loading into a torpedo tube. In the latter case they would be inserted into a torpedo tube before final launch verification checks are made.

Air-Launched Weapon

Strike weapons that are launched from land-based aircraft are brought out of storage and removed from containers ("decanned") as appropriate. They are inspected and subjected to a verification test to ensure that they are indeed ready for use, and may undergo final assembly steps at that point. The weapons are then moved to their launch aircraft on specialized pieces of handling equipment, at which point they are uploaded on the aircraft. Any required mission data may be loaded into the weapon either before the aircraft departs the base or during the flight to the target area.

The strike aircraft takes off from the air base and flies to the launch point. The flight profile can be variable in terms of time, altitude, speed, and maneuvers, and may expose the weapon to a variety of weather conditions. Sometimes the mission is aborted before the weapon is launched, resulting in the weapon being brought back to base. Download from the aircraft and return to user storage may occur under those circumstances. The weapon may be exposed to several airborne mission cycles before it is actually expended.

Weapons deployed from aircraft carriers follow a generally similar path, although the environmental conditions on board the aircraft carrier differ somewhat from typical land air bases. Electromagnetic radiation, for example, tends to be higher on board a carrier due to the very close proximity of the radar antennas to the flight deck. Figure B.2 may help the reader to visualize the dramatic physical difference between the two basing alternatives. All of the basic functions of a tactical airfield must be accommodated on a vessel that is a small fraction of the size of the land base. Carrier-based weapons must also tolerate the shock and structural loads associated with catapult takeoffs and arrested landings.

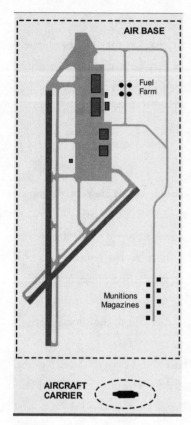

Figure B.2 Size Comparison: Air Base and Aircraft Carrier

WEAPON LAUNCH AND FLYOUT

Weapons vary in terms of their need for prelaunch warm-up or guidance initialization. Those that require it need a little time to satisfy this step just before the weapon is launched. When a launch command is issued, generally by a button push or a trigger pull, the different types of strike weapons take different paths in departing their launch platforms. In point of fact, each specific weapon has a tailored environmental-design profile associated with its unique carriage, launch, and flyout characteristics. However, for purposes of this discussion, we will take a more generalized approach to describing basic environmental inputs.

For ship and submarine launches, there is some level of environmental stimulus from the launcher itself, but this is largely overshadowed by hydrodynamic forces (where applicable) and rocket motor boost. The boost phase is generally short but does introduce substantial longitudinal accelerations as well as acoustic, shock, and vibration inputs. When the booster has been expended, the primary propulsion motor is activated, which then powers the weapon for the remainder of the flight. This may be either a second rocket motor or an air-breathing motor of some type (e.g., turbojet, turbofan, ramjet). This engine also introduces forces, vibrations, and noise to the design environment.

The flyout of a contemporary ship- or submarine-launched strike weapon is characterized by an extended period of flight at a near-constant velocity and altitude, called the midcourse phase, followed by maneuvering and possible speed and altitude changes as the weapon approaches the target, called the terminal phase. The conditions vary with the different weapon types, resulting in design criteria tailored to the specific weapon. For example, a low-altitude cruise missile might experience flight at a few hundred feet above sea level for many miles at a subsonic Mach number. Alternatively, a high-speed weapon might accelerate to a supersonic Mach number while climbing to a high altitude, cruise at that condition for a substantial period, and then dive steeply on its target in the terminal phase. The flight loads and aerodynamic heating resulting from these two contrasting trajectories would differ markedly.

Air-launched weapons can include ballistic, glide, or powered types. At launch, ballistic and glide weapons are generally pushed away from their captive carriage positions by pistons that are part of the bomb rack, the mechanism that holds them in place on the aircraft. The ejection sequence imparts a shock to the weapon as well as a downward force.

Powered air-launched weapons may also be launched as just described, with the propulsion system ignition delayed briefly to ensure that the weapon has cleared the aircraft before it activates. Alternatively, the weapon may be launched from a device that allows it to be powered and moving forward as it leaves the aircraft; this typically applies to smaller, rocket-propelled systems.

Once launched, ballistic weapons fall away from the aircraft under the influence of gravity and atmospheric conditions. Guided bombs and glide weapons experience maneuver loads as well. Powered weapons will be exposed to conditions similar to that described above for ship- and submarine-launched weapons during the flyout period.

Ballistic weapons, by definition, do not maneuver. However, all other forms of strike weapons are expected to maneuver in the terminal phase of flight as the guidance system attempts to place the weapon on target. The extent of such maneuvering is unique to each weapon.

A few strike weapons dispense submunitions in the terminal phase. To place the submunitions on the target, the dispense position must be above and short of the actual target. The dispensing event generally introduces substantial shock to the system as the weapon skin is explosively opened, allowing the submunitions to disperse.

Most weapons, however, are intended to actually hit the target. If they are expected to penetrate into or partway into the target, the warhead section must be designed to survive a massive impact shock and extremely high forces before it is detonated.

All of these myriad postlaunch conditions lead to a challenging array of environmental-design conditions, many of them unique to the specific weapon system.

SPECIFIC DESIGN CONDITIONS

As should be obvious at this point, the definition of the environmental-design criteria for any particular strike weapon is a nontrivial task that is addressed by specialists who make use of field measurements and analyses of a wide variety of settings and conditions. The specific criteria are tailored to the expected life cycle profile of the hardware being developed, and are intended to encompass realistic combinations of conditions without exaggerating or "overdesigning" the environmental inputs. During the tailoring process, estimates are made of the amount of time and the number of repetitions that the hardware might be exposed to the particular environment. Indications are also provided as to the condition of the hardware while it experiences the input: For example, is it "bare" or inside a container? Is it operating, or is it in a nonoperating state? Many environmental parameters, especially those involving vibration, shock, and acoustics, become highly technical when described in engineering design terms.

Appendix C

Military Designations and Weapons Names

This appendix provides some insights into the manner in which American strike weapons are identified. Other nations use designation practices that differ from the U.S. system; those practices are not addressed.

Numbered Designations

Guided Missiles

The U.S. Department of Defense uses a structured and standardized system to establish formal designations for strike weapons, and, indeed, for all guided weaponry. Such devices are identified with a series of letters and numbers. Such examples as "AGM-65E," "CATM-84A," "RGM-84D," or "BGM-109G" can be translated by breaking down the designation into several parts:

First letter: status prefix (may not apply to all devices)
Second letter: launch environment (always appears)
Third letter: basic mission (always appears)
Fourth letter: vehicle type (always appears)
Numbers: design number, issued in numerical sequence (always appears)
Final letter: version of the basic design (may not always appear)

The letters used in the several parts of the designator are defined by the Department of Defense as indicated below. Given the wide range of historic, current, and potential guided-weapon systems, both nuclear (strategic) and nonnuclear (tactical), there is considerable breadth in the possible combinations of letters.

STATUS PREFIX
C: captive
D: dummy
J: special test (temporary)
N: special test (permanent)
X: experimental
Y: prototype
Z: planning

LAUNCH ENVIRONMENT
A: air
B: multiple
C: coffin (e.g., Atlas ICBM)
F: individual (e.g., shoulder-fired missile)
G: surface (ground)
H: silo stored
L: silo launched (e.g., Minuteman ICBM)
M: mobile
P: soft pad
R: ship
S: space
U: underwater

BASIC MISSION
C: transport
D: decoy
E: electronic/communications
G: surface (ground) attack
I: aerial/space intercept
L: launch detection/surveillance
M: scientific/calibration
N: navigation

Q: drone

S: space support

T: training

U: underwater attack

W: weather

VEHICLE TYPE

B: booster

M: guided missile

N: probe

R: rocket

S: satellite

Thus "AGM-65E" would be translated as an "air-launched surface attack guided missile, design series number 65, fifth (E) version." A "CATM-84A" would be translated as a "captive air-training missile, design series number 84, first (A) version."

The numbering sequence began many years back and continues in strict numerical order as new systems are developed. To avoid possible identification errors, there is only one numbering sequence for all U.S. guided weapons. This prevents duplication of numbers for different launch or mission types. For example, the number 65 is only assigned to the AGM-65 family; there is no AIM-65, RIM-65, UUM-65, etc.

MUNITIONS

Unguided weapons such as bombs and dispensers, as well as certain guided but unpowered weapons, use a somewhat different designation system as described below. This again uses letters and numbers in an arrangement that is visually similar to the guided-missile designation system, but different meanings are assigned to the characters. The scope of the munitions designation system is quite large, encompassing everything from guided munitions and unguided bombs to gun pods, internal guns, leaflet dispensers, bomb racks, and so on. Rather than overly complicate this appendix, only those items that relate to strike weaponry will be described here.

The munitions designation system begins with a two-letter functional designator, followed by the letter "U" to indicate "unit," which is then followed by a design series number, and ends with an additional pair of letters that indicate the version and installation type, as in "CBU-87A/B." The functional designators of interest here include:

BL: bomb or bomblet
BD: simulated bomb
CB: cluster bomb
GB: guided bomb

The first letter following the numbers in the designation indicate the version of the design in the same manner as used in the guided-missile designators. Thus "A" would indicate the first version, "B" would indicate the second version, and so on. The letter following the slash mark indicates the type of installation:

A: fixed installation on aircraft (e.g., an internal gun)
B: an expendable item released or launched from an aircraft

The example "CBU-87A/B" would therefore be interpreted as a cluster bomb, design series 87, first (A) version, intended to be released from an aircraft. It is relatively common when describing a munition in general terms to omit the letters and slash following the number series; in those cases, "CBU-87A/B" would simply become "CBU-87."

Other Designators

There are a few, mostly older items of strike weaponry that use earlier designations. These would include the family of low-drag, general-purpose bombs identified as Mk 81, Mk 82, Mk 83, and Mk 84. Other examples include the M117 and M118 bombs as well as the Mk 20 "Rockeye" antiarmor cluster weapon. Especially confusing is the "Walleye" family of TV-guided glide weapons, which was originally designated AGM-62 but which was redesignated by the Navy using a Mark (Mk) and Modification (Mod) series. It began with Guided Weapon, Mk 1 Mod 0, which designated the original Walleye I (small weapon, small wing) variant. Over time, with the development of the larger Walleye II weapon and the use of larger wings, a data link, and updated guidance components, the series eventually reached Guided Weapon, Mk 27. Walleye is no longer in service use, so the designation issue does not impact current activities. However, for purposes of historical context, the Walleye family is identified in this book by AGM-62, its original designator.

Names

The formal designations are generally supplemented with official or unofficial weapon system names, especially for guided systems. There is a formal process that is followed to obtain an official name for a new weapon, but that process is often redundant when an established program acronym, such as JSOW or JDAM, is adopted as the system name. Other weapons never seem to receive a name, official or otherwise.

Weapons List

This appendix ends with a listing of U.S. nonnuclear strike weapons, including a number of obsolete items shown for historical context. Listed first (table C.1) are guided missiles, presented in numerical order by their design series number. A munitions listing then follows (table C.2), with guided munitions, unguided munitions, and submunitions listed alphabetically. Official names as well as unofficial (popular) names are attached where known, along with clarifying notes as appropriate. (Weapons characteristics are found in appendix D.)

At first glance the length of the list may give the impression that there is an abundance of weaponry available to combat units. The munitions listing will be particularly prone to such an assumption. In reality the choices and quantity of strike weapons available in the field has limits. You will note in the listing that some items are no longer in service and others are unique to a particular military service. Furthermore, some items are tailored to specific launch platforms due to their size, weight, or application. As a result the actual "catalog" of strike resources tends to shrink rather substantially at the operational level.

Designation	Name	Notes
AGM-12	Bullpup	Command guided; no longer in service
AGM-45	Shrike	ARM; no longer in service
AGM-53	Condor	Canceled; TV seeker, data link, rocket motor
AGM-62:	Walleye	TV guided; no longer in service
Mk 1	Walleye I	Small LSC warhead, small wings
Mk 21	Walleye I ERDL	Big wings, data link
Mk 5	Walleye II	Large LSC warhead, small wings
Mk 23	Walleye II ERDL	Big wings, data link
AGM-65:	Maverick	
AGM-65A, B	TV Maverick	USAF; TV seeker, CSC warhead
AGM-65C	Laser Maverick	USAF; laser seeker, CSC warhead; canceled
AGM-65D	IR Maverick	USAF; IIR seeker, CSC warhead
AGM-65E	Laser Maverick	USMC; laser seeker, large warhead
AGM-65F	IR Maverick	USN; IIR seeker, large warhead
AGM-65G	IR Maverick	USAF; IIR seeker, large warhead
AGM-78	Standard ARM	ARM; no longer in service
AGM-83	Bulldog	Laser guided; canceled
AGM-84:	Harpoon	
AGM-84A–D	Harpoon	Antiship; active radar guidance, air launch
AGM-84E	SLAM	USN; land attack, IIR seeker
AGM-84H	SLAM-ER	USN; land attack, upgraded SLAM
RGM-84	Harpoon	USN; antiship; active radar guidance, ship/sub launch
AGM-86C	C-ALCM	USAF; nonnuclear ALCM
AGM-88:	HARM	
AGM-88A–D	HARM	ARM
AGM-88E	AARGM	Advanced variant with dual mode seeker
BGM-109	TLAM	USN; nonnuclear Tomahawk
AGM-114	Hellfire	Laser guided, helicopter or UAV launch
AGM-119	Penguin	IR-guided antiship, helicopter launch; Norwegian design
AGM-122	SideARM	USMC; ARM variant of Sidewinder
AGM-123	Skipper II	Powered LGB (Shrike motor); no longer in service
AGM-130		USAF; powered GBU-15 (underslung rocket motor)
AGM-136	Tacit Rainbow	Loitering ARM; canceled
AGM-137	TSSAM	Canceled/never produced; stealthy design
AGM-142	Have Nap/Raptor	USAF; TV guided, Israeli design
AGM-154:	JSOW	
AGM-154A	JSOW CEB	GPS guided, BLU-97 dispenser
AGM-154B	JSOW BLU-108	GPS guided, BLU-108 dispenser; canceled
AGM-154C	JSOW Unitary	GPS + IIR; unitary warhead
AGM-158	JASSM	Stealthy tactical cruise missile

Table C.1 U.S. Strike Weapons: Guided Missiles

Designation	Name	Notes
BLU-77		Used in APAM; antiarmor submunition
BLU-97	CEB	Used in CEM & JSOW; multipurpose submunition
BLU-108		Used in SFW; antiarmor submunition
BLU-109		2,000-lb-class penetrator warhead
BLU-126	LCDB	500-lb-class Low Collateral Damage Bomb
CBU-55	Low-speed FAE	3-canister FAE weapon; no longer in service
CBU-59	APAM	Cluster weapon
CBU-72	High-speed FAE	3-canister FAE weapon; no longer in service
CBU-78	Gator	Mine dispenser
CBU-87	CEM	Combined-effects munition (dispenser)
CBU-89	Gator	Mine dispenser
CBU-97	SFW	USAF; sensor-fuzed weapon (antiarmor)
GBU-8	HOBOS	USAF; no longer in service
GBU-10		Mk 84 LGB
GBU-12		Mk 82 LGB
GBU-15		USAF; TV or IIR guided, data link
GBU-16		Mk 83 LGB
GBU-20		Canceled
GBU-22	Paveway III LLLGB	Mk 82 warhead
GBU-24	Paveway III LLLGB	Mk 84 or BLU-109 warhead
GBU-27		USAF; special configuration LGB for F-117
GBU-28		USAF; deep-penetrator LGB; very heavy
GBU-29	JDAM	GPS guided; 2,000-lb-class warhead; canceled
GBU-30	JDAM	GPS guided; 1,000-lb-class warhead; canceled
GBU-31	JDAM	GPS guided; Mk 84 or BLU-109 warhead
GBU-32	JDAM	GPS guided; Mk 83 warhead
GBU-35	JDAM	GPS guided; 1,000-lb-class warhead
GBU-37	GAM	USAF; GPS-guided heavy-penetrator warhead
GBU-38	JDAM	GPS guided; Mk 82 or BLU-126 warhead
GBU-39	SDB	GPS-guided small-diameter bomb (penetrator)
GBU-40	SDB	GPS-guided small-diameter bomb (blast-frag)
GBU-43	MOAB	USAF; GPS-guided very heavy blast weapon
GBU-53	SDB-II	SDB with wings, trimode seeker (IIR, mmw, laser)
GBU-54	Laser JDAM	GPS + laser, 500-lb-class warhead
GBU-55	Laser JDAM	GPS + laser, 1,000-lb-class warhead
GBU-56	Laser JDAM	GPS + laser, 2,000-lb-class warhead
GBU-57	MOP	USAF; GPS-guided very heavy penetrator
M117		750-lb-class bomb
M118		3,000-lb-class bomb
Mk 20	Rockeye	Antiarmor cluster weapon
Mk 81		250-lb-class low-drag bomb
Mk 82		500-lb-class low-drag bomb
Mk 83		1,000-lb-class low-drag bomb
Mk 84		2,000-lb-class low-drag bomb

Table C.2 U.S. Strike Weapons: Munitions

Appendix D

Strike Weapons Characteristics

This appendix provides summary data on a number of U.S. strike weapons, both past and present. All information was drawn from unclassified sources and therefore cannot be taken as official or authoritative. It is presented in the interests of illustrating the general physical and functional characteristics of this category of military ordnance.

Tabulated Data

Tables D.1a and D.1b list basic characteristics of representative strike weapons, starting first with sea-launched weapons (i.e., Harpoon and Tomahawk) and then encompassing the broad array of air-launched weapons (ballistic, guided bombs, glide weapons, ARMs, and powered weapons). The table identifies the weapon and then lists the type of propulsion employed (if any), the type of warhead, and the principal type of guidance used. Physical characteristics include overall length (L), maximum body diameter (Dmax), maximum wing span (bmax), and gross weight (Wgross).

A check mark ($\sqrt{}$) immediately to the left of the weapon designation indicates that the item is not in service. Blanks in the table indicate that data is not available in open literature.

The following abbreviations are used in these tables.

General

@: number of submunitions carried

in: inch(es)	LD: low drag configuration
lb: pound(s)	NC: body cross section is not circular

WARHEAD TYPES

CSC: conical-shaped charge

frag: fragmentation

GP: general purpose

LSC: linear-shaped charge

pen.: penetrator

GUIDANCE TYPES

ARM: antiradiation

DL: data link

IIR: imaging infrared

lsr: laser

mmw: millimeter wave

TV: television

Table D.2 separates the weapons into broad standoff categories, providing a general notion of their maximum-range capabilities. Standoff is often heavily influenced by launch conditions (i.e., altitude and speed at launch), but this refinement is not depicted here.

Plotted Data

Figures D.1 through D.3 illustrate certain physical parameters that may help place the weapons in relative perspective. Figure D.1 organizes the weapons by overall length, from the shortest to the longest. (The extremely long GBU-43 massive ordnance air blast, or MOAB, is not depicted.) In the case of the ship-launched weapons, Harpoon and Tomahawk, a shaded section at the right end of the bar indicates the length of the rocket booster motor, which is jettisoned after it accelerates the missile to flight speed.

Length is one of several physical parameters that determines the compatibility of a weapon with any given launch platform. Very long weapons may only be usable on very large aircraft or with large shipboard launchers. Length is also is a factor in weapon handling and storage. This is especially true on board aircraft carriers, where weapons elevators and transfer spaces place serious constraints on long weapons.

Figure D.2 organizes the weapons by gross weight, from lightest to heaviest. (Not illustrated are the extremely heavy GBU-37 GPS-aided munition

[GAM], GBU-28, GBU-43 MOAB, and GBU-57 massive ordnance penetrator [MOP].) The shaded portion of the bar indicates the weight of the weapon payload, the warhead. The weight of the rocket booster motor used on the ship-launched Harpoon and Tomahawk is indicated with a vertical tick mark on the right end of the bar.

Weight is another important compatibility parameter, with very heavy weapons only usable from very large aircraft or large surface launchers.

Figure D.3 illustrates the weapons' "payload fraction," which is the ratio of the warhead weight to the gross weight. In one sense this can be taken as the relative delivery efficiency of the weapon; the higher the fraction, the more efficient the weapon is at delivering ordnance on target. However, the desire for a large payload fraction is often offset by the need for standoff, which requires that some of the total weapon weight must be devoted to propulsion and lifting elements. This trade-off can be observed by noting the difference in payload fraction between the ballistic Mk 82 bomb, the AGM-154 JSOW glide weapon, and the AGM-84A Harpoon, which all carry a nominal 500-pound payload but to different standoff ranges. The effects of packaging efficiency also become apparent when comparing ballistic dispensers, such as the CBU-87 CEM, with ballistic bombs, such as the Mk 80 series.

	Designation	Name	Propulsion	Warhead	Guidance	L (in)	Dmax (in)	bmax (in)	Wgross (lb)
Sea Launch									
	RGM-84	Harpoon		Pen.-blast	Radar		13.5		
		At launch	Rocket			183			1,543
		Booster gone	Turbojet			152		36	1,185
	BGM-109	TLAM (Tomahawk)		Pen.-blast	Various		20.3		
		At launch	Rocket			244			3,370
		Booster gone	Turbofan			218		103.6	2,270
Air Launch									
Ballistic Weapons									
	Mk 81 (LD)		None	Gen'l purpose	None	72	9	12.6	260
	Mk 82 (LD)		None	Gen'l purpose	None	91	10.75	15.1	510
	Mk 83 (LD)		None	Gen'l purpose	None	119	14	19.6	985
	Mk 84 (LD)		None	Gen'l purpose	None	151	18	25.3	2,030
	Mk 20	Rockeye	None	247 @ Mk 118	None	92	13.2	34.4	510
	CBU-87	CEM	None	202 @ BLU-97	None	92	15.6	42	950
Guided Bombs									
	GBU-10	LGB	None	Mk 84 GP	Laser	170	18	65.6	2,083
	GBU-12	LGB	None	Mk 82 GP	Laser	131	10.75	52	611
	GBU-16	LGB	None	Mk 83 GP	Laser	145	14	63.6	1,092
√	GBU-22	LLLGB	None	Mk 82 GP	Laser	~138	10.75		~700
	GBU-24	LLLGB	None	Mk 84 GP	Laser	173	18	81	2,354
	GBU-24	LLLGB	None	BLU-109 pen.	Laser	170	16	81	2,315
	GBU-27	LLLGB	None	BLU-109 pen.	Laser	167	16	66	2,170
	GBU-28	LLLGB	None	BLU-109 pen.	Laser	229	14.5	66	4,676
√	GBU-29	JDAM	None	Mk 81 GP	GPS				
√	GBU-30	JDAM	None	Mk 82 GP	GPS				
	GBU-31	JDAM	None	Mk 84 GP	GPS	153	18	25.3	2,150
	GBU-31	JDAM	None	BLU-109 pen.	GPS	148	16	25.3	2,126
	GBU-32	JDAM	None	Mk 83 GP	GPS	119	14	19.6	1,034
	GBU-35	JDAM	None	BLU-110	GPS	119	14	19.6	1,034
√	GBU-36	GAM	None	Mk 84 GP	GPS				
	GBU-37	GAM	None	BLU-113 pen.	GPS	210	14.5		4,500
	GBU-38	JDAM	None	Mk 82 GP	GPS	92.6	10.75	17	558
	GBU-39	SDB	None	Penetrator	GPS	71	7.5		285
	GBU-43	MOAB	None	Blast	GPS	362	40.5		22,600
	GBU-54	Laser JDAM	None	Mk 82 GP	GPS + laser				
√	GBU-55	Laser JDAM	None	Mk 83 GP	GPS + laser				
√	GBU-56	Laser JDAM	None	Mk 84 GP	GPS + laser				
	GBU-57	MOP	None	Penetrator	GPS	240	31.5		30,000

Table D.1a Strike Weapons Data

Designation	Name	Propulsion	Warhead	Guidance	L (in)	Dmax (in)	bmax (in)	Wgross (lb)
Glide Weapons								
√ AGM-62/Mk 1	Walleye I	None	LSC	TV	136	15	45	1,120
√ AGM-62/Mk 21	Walleye I ERDL	None	LSC	TV + DL	136	15	54.3	1,200
√ AGM-62/Mk 5	Walleye II	None	Large LSC	TV	159	18	51.1	2,455
√ AGM-62/Mk 23	Walleye II ERDL	None	Large LSC	TV + DL	159	18	68.1	2,490
GBU-15		None	Mk 84 GP	TV/IIR	155	18	59	2,500
AGM-154A	JSOW	None	145 @ BLU-97	GPS	160	13.3x17.4	106	1,045
√ AGM-154B	JSOW	None	6 @ BLU-108	GPS	160	13.3x17.4	106	1,045
AGM-154C	JSOW	None	Pen.-blast	GPS + IIR	160	13.3x17.4	106	1,045
GBU-40	SDB-II	None	Pen.-blast	GPS + IIR	71	7.5	54	285
GBU-53	SDB-II	None	Pen.-blast	GPS, IIR, mmw, lsr				
Antiradiation Missiles (ARMs)								
√ AGM-45	Shrike	Rocket	Frag	ARM	121	8	36.3	426
√ AGM-78	Standard ARM	Rocket	Frag	ARM	180	13.5	36	1,356
AGM-88A–D	HARM	Rocket	Frag	ARM	164	10	44	805
AGM-88E	AARGM	Rocket	Frag	Dual mode	164	10	44	805
AGM-122	SideARM	Rocket	Frag	ARM	118	5	24.8	200
√ AGM-136	Tacit Rainbow	Turbojet	Frag	ARM	100	NC	62	430
Powered Strike Weapons								
√ AGM-12B	Bullpup	Rocket	Pen.-blast	Command	126	12	37	571
√ AGM-12C	Bullpup	Rocket	Pen.-blast	Command	161	18	48	1,790
√ AGM-53	Condor	Rocket	LSC	TV + DL	166	17	53	2,100
AGM-65A, B	Maverick	Rocket	CSC	TV	98	12	28.5	462
√ AGM-65C	Maverick	Rocket	CSC	Laser	98	12	28.5	462
AGM-65D	Maverick	Rocket	CSC	IIR	98	12	28.5	485
AGM-65E	Maverick	Rocket	Pen.-blast	Laser	98	12	28.5	667
AGM-65F	Maverick	Rocket	Pen.-blast	IIR	98	12	28.5	667
AGM-65G	Maverick	Rocket	Pen.-blast	IIR	98	12	28.5	667
√ AGM-83	Bulldog	Rocket	Pen.-blast	Laser	126	12	37	620
AGM-84A–D	Harpoon	Turbojet	Pen.-blast	Radar	152	13.5	36	1,185
AGM-84E	SLAM	Turbojet	Pen.-blast	IIR	175	13.5	36	1,400
AGM-84H	SLAM-ER	Turbojet	Pen.-blast	IIR	172	13.5	95.6	1,600
AGM-86C	C-ALCM	Turbofan	Blast-frag	GPS	249	NC	143.6	3,250
AGM-114	Hellfire	Rocket	CSC	Laser	64	7	12.8	99
AGM-119	Penguin	Rocket	Pen.-blast	IR	118	11	55	850
√ AGM-123	Skipper II	Rocket	Gen'l purpose	Laser	172	14	63.6	1,285
AGM-130		Rocket	Pen.-blast	GPS + IIR	156	18 + 9	59	2,980
√ AGM-137	TSSAM	Turbofan	Pen.-blast	GPS + IIR	168	NC	100	2,000
AGM-142	Have Nap	Rocket	Pen.-blast	IIR + DL	191	21	78	3,020
AGM-158	JASSM	Turbofan	Pen.-blast	GPS + IIR + DL	168	NC	105	2,250
Submunitions								
BLU-97	CEB	None	CSC + frag	None	6.6	2.51	...	3.4
Mk 118	Rockeye	None	CSC	None	11	1.91	...	1.32

Table D.1b Strike Weapons Data

Standoff Class				Launch Platform			Primary Targets			Weapon
SOTD	SOAD	SOPD	CI	Ship	Sub	Aircraft	Ships	Radars	Land Targets	
X						X			X	AGM-86C C-ALCM
X				X	X				X	BGM-109 Tomahawk TLAM-C
	X			X	X	X	X			RGM-84A/AGM-84A Harpoon
	X					X	o		X	AGM-84E SLAM/AGM-84H SLAM-ER
	X					X	o	X		AGM-88 HARM/AARGM
	X					X	o		X	AGM-142 Have Nap
	X					X			X	AGM-158 JASSM
		X				X	X			AGM-119 Penguin
		X				X			X	AGM-130
		X				X	o		X	AGM-154 JSOW
			X			X	o		X	AGM-65 Maverick
			X			X			X	AGM-114 Hellfire
			X			X	o		X	GBU-10, -12, -16 LGB
			X			X			X	GBU-15
			X			X	o		X	GBU-22, -24 LLLGB
			X			X	o		X	GBU-31, -32, -35 JDAM
			X			X			X	GBU-39 SDB
			X			X			X	GBU-54, -55, -56 Laser JDAM
			X			X	o		X	Dispensers (CBU-87 CEM, Mk 20 Rockeye)
			X			X			X	GP bombs (Mk 80 family, etc.)
			X			X			X	Special large bombs (MOP, MOAB, etc.)

SOTD = standoff outside theater defenses (very long range)
SOAD = standoff outside area defenses (long range)
SOPD = standoff outside point defenses (medium range)
CI = close-in (short range)

X = principal mission
o = secondary mission

Note: Radar targets may be on ships or on land.

Table D.2 Weapon Standoff and Application

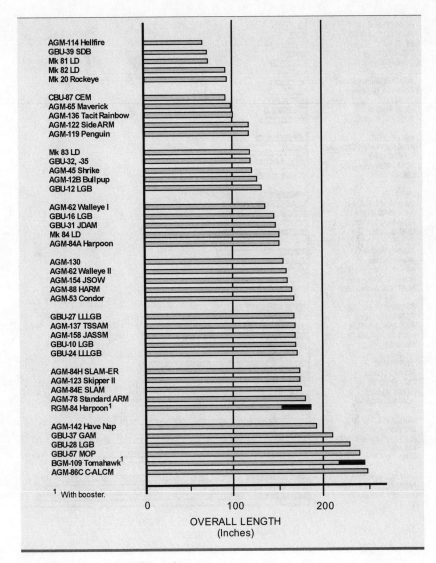

AGM-114 Hellfire
GBU-39 SDB
Mk 81 LD
Mk 82 LD
Mk 20 Rockeye

CBU-87 CEM
AGM-65 Maverick
AGM-136 Tacit Rainbow
AGM-122 SideARM
AGM-119 Penguin

Mk 83 LD
GBU-32, -35
AGM-45 Shrike
AGM-12B Bullpup
GBU-12 LGB

AGM-62 Walleye I
GBU-16 LGB
GBU-31 JDAM
Mk 84 LD
AGM-84A Harpoon

AGM-130
AGM-62 Walleye II
AGM-154 JSOW
AGM-88 HARM
AGM-53 Condor

GBU-27 LLLGB
AGM-137 TSSAM
AGM-158 JASSM
GBU-10 LGB
GBU-24 LLLGB

AGM-84H SLAM-ER
AGM-123 Skipper II
AGM-84E SLAM
AGM-78 Standard ARM
RGM-84 Harpoon[1]

AGM-142 Have Nap
GBU-37 GAM
GBU-28 LGB
GBU-57 MOP
BGM-109 Tomahawk[1]
AGM-86C C-ALCM

[1] With booster.

0   100   200

OVERALL LENGTH
(Inches)

Figure D.1 Weapon Overall Length

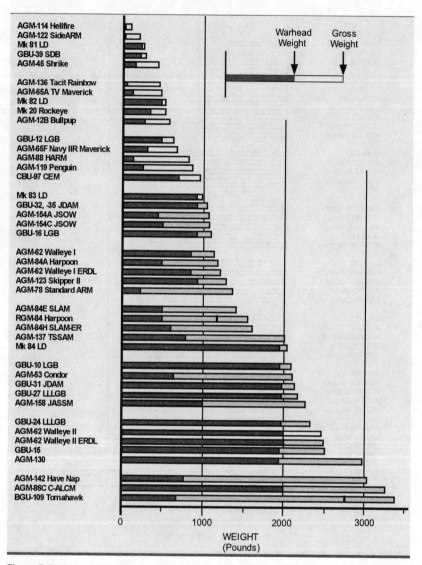

Figure D.2 Weapon Weight

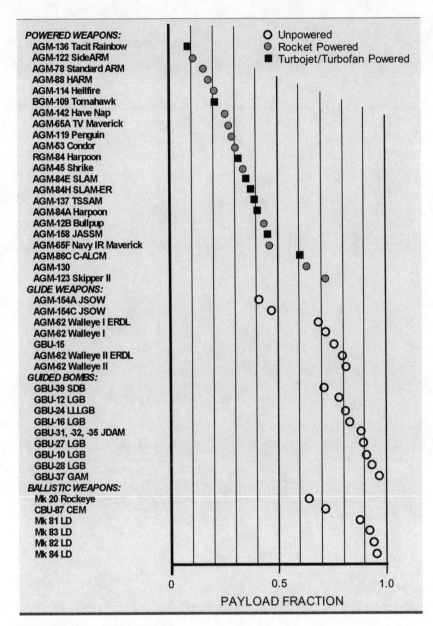

Figure D.3 Weapon Payload Fraction

Appendix E

Sources of Additional Information

Web sites

Internet search engines can always be used to locate information on specific weapons of interest. The best results are achieved when using the system designation rather than the popular name or acronym (e.g., search for "AGM-65" rather than for "Maverick"). The website list that follows focuses on more generalized weapons information and collections of strike weapons data, and was current as of September 2011.

Government Sites

U.S. AIR FORCE
- Air Force Information Office
    http://www.af.mil/information/factsheets/index.asp
- Air Force Museum
    http://www.nationalmuseum.af.mil

U.S. NAVY
- Naval Air Systems Command
    http://www.navair.navy.mil/index.cfm
- Naval Air Warfare Center, Weapons Division
    http://www.navair.navy.mil/nawcwd
- Navy Information Office
    http://www.navy.mil/navydata/infoIndex.asp?id=A

Nongovernment Sites

- Aviation Week
  http://www.aviationweek.com/aw/
- China Lake Weapons Museum
  http://www.chinalakemuseum.org
- Eglin Air Force Base Armament Museum
  http://www.afarmamentmuseum.com
- Federation of American Scientists
  http://www.fas.org/programs/ssp/man/index.html
- Global Security, Military Systems
  http://www.globalsecurity.org/military/systems/index.html
- Jane's Defence Systems
  http://www.janes.com/products/janes/defence/

Books

Chant, Christopher. *Aircraft Armaments Recognition*. Shepperton, Surrey, UK: Ian Allan, 1989.

Dougherty, Martin. *Modern Air-Launched Weapons*. New York: Metro Books, 2010.

Flack, Jeremy. *NATO Air-Launched Weapons*. Ramsbury, Marlborough, UK: Crowood Press, 2002.

Friedman, Norman. *The Naval Institute Guide to World Naval Weapon Systems*. Annapolis, MD: Naval Institute Press, 2006.

Gunston, Bill. *The Illustrated Encyclopedia of Aircraft Armament*. London: Salamander Books, 1988.

———. *The Illustrated Encyclopedia of the World's Rockets & Missiles*. London: Salamander Books, 1979.

———. *An Illustrated Guide to Modern Airborne Missiles*. New York: Arco, 1983.

Lee, R. G., et al. *Guided Weapons*. McLean, VA: Pergamon-Brassey's, 1983.

Lennox, Duncan, and Arthur Rees, eds. *Jane's Air-Launched Weapons*. Various updated eds. Coulsdon, Surrey, UK: Jane's Information Group, 1990–.

Polmar, Norman. *The Naval Institute Guide to the Ships and Aircraft of the U.S. Fleet*. Various updated eds. Annapolis, MD: Naval Institute Press, 1997–.

Pretty, Ronald. *Jane's Pocket Book of Missiles*. New York: Macmillan, 1975.

Pretty, R. T., ed. *Jane's Weapon Systems*. Various updated eds. London: Jane's
　　Publishing, 1985–.

Walker, J. R. *Air-to-Ground Operations*. McLean, VA: Pergamon-Brassey's, 1987.

Limited-Distribution Publications

Heaston, R. J., and C. W. Smoots. *Introduction to Precision Guide Munitions*.
　　GACIAC HB-83-01. Chicago: IIT Research Institute, 1983.

Naval Weapons Handbook. NAWCWD TP 8451. China Lake, CA: Naval Air
　　Warfare Center–Weapons Division, 1999. (Distribution limited to U.S.
　　government and associated contractors.)

1990 Weapons File. Eglin, FL: Munitions Systems Division, Eglin AFB, 1990.
　　(Distribution limited to U.S. government and associated contractors.)

The World's Missile Systems. Pomona, CA: General Dynamics, 1988.

Index

About the Author

Mr. Knutsen was a part of the strike weapons development community for three decades, serving primarily at the U.S. Navy's airborne weapons establishment at China Lake, California. He is perhaps best known for his association with such programs as Walleye, Harpoon, and JSOW. He now makes his home in the forests of northern California.

The Naval Institute Press is the book-publishing arm of the U.S. Naval Institute, a private, nonprofit, membership society for sea service professionals and others who share an interest in naval and maritime affairs. Established in 1873 at the U.S. Naval Academy in Annapolis, Maryland, where its offices remain today, the Naval Institute has members worldwide.

Members of the Naval Institute support the education programs of the society and receive the influential monthly magazine *Proceedings* or the colorful bimonthly magazine *Naval History* and discounts on fine nautical prints and on ship and aircraft photos. They also have access to the transcripts of the Institute's Oral History Program and get discounted admission to any of the Institute-sponsored seminars offered around the country.

The Naval Institute's book-publishing program, begun in 1898 with basic guides to naval practices, has broadened its scope to include books of more general interest. Now the Naval Institute Press publishes about seventy titles each year, ranging from how-to books on boating and navigation to battle histories, biographies, ship and aircraft guides, and novels. Institute members receive significant discounts on the more than eight hundred Press books in print.

Full-time students are eligible for special half-price membership rates. Life memberships are also available.

For a free catalog describing Naval Institute Press books currently available, and for further information about joining the U.S. Naval Institute, please write to:

Member Services
U.S. NAVAL INSTITUTE
291 Wood Road
Annapolis, MD 21402-5034
Telephone: (800) 233-8764
Fax: (410) 571-1703
Web address: www.usni.org